WJEC
Mathematics
for A2 Level

Pure & Applied Revision Guide

Stephen Doyle

Illuminate Publishing

Published in 2021 by Illuminate Publishing Limited, an imprint of Hodder Education, an Hachette UK Company, Carmelite House, 50 Victoria Embankment, London EC4Y 0DZ

Orders: please contact Hachette UK Distribution, Hely Hutchinson Centre, Milton Road, Didcot, Oxfordshire, OX11 7HH. Telephone: +44 (0)1235 827827. Email: education@hachette.co.uk. Lines are open from 9 a.m. to 5 p.m., Monday to Friday. You can also order through our website: www.hoddereducation.co.uk

British Library Cataloguing in Publication Data

A catalogue record for this book is available from the British Library

ISBN 978-1-912820-34-4

Printed by Ashford Colour Press, UK

Impression 4

Year 2024

Hachette UK's policy is to use papers that are natural, renewable and recyclable products and made from wood grown in well-managed forests and other controlled sources. The logging and manufacturing processes are expected to conform to the environment regulations of the country of origin.

Editor: Geoff Tuttle
Cover design: Neil Sutton
Text design: GreenGate Publishing Services, Tonbridge, Kent
Layout: Kamae Design

Photo credits

Cover: Klavdiya Krinichnaya/Shutterstock; **p15** Radachynskyi Serhii/Shutterstock; **p20** Anatoli Styf/ Shutterstock; **p32** Jeeraphun Kulpetjira/Shutterstock; **p42** Jan Miko/Shutterstock; **p53** Africa Studio/ Shutterstock; **p62** Fotos593/Shutterstock; **p69** jo Crebbin/Shutterstock; **p81** Jose Antonio Perez/ Shutterstock; **p90** Rena Schild/Shutterstock; **p99** Marco Ossino/Shutterstock; **p116** Sergey Nivens/ Shutterstock; **p131** Blackregis/Shutterstock; **p139** urickung/Shutterstock; **p145** Gajus/Shutterstock; **p154** Will Rodrigues/Shutterstock; **p162** Andrea Danti/Shutterstock; **p169** sirtravelalot/Shutterstock.

Acknowledgements

The author and publisher wish to thank Sam Hartburn for her careful attention when reading this book.

Contents

How to use this book

The purpose of this book is to aid revision just before the examination, so it is assumed you have completed the course and have a reasonable knowledge of the topics. The book has been specifically written for the WJEC A2 Level course and has been produced by an experienced author and teacher. The book includes full coverage of both the Pure and Applied Mathematics units, so it covers the A2 course.

A feature of the new exams is the lack of scaffolding in some questions. When questions have scaffolding, they guide you through the steps by questioning you on each step as you go along until you arrive at a final answer. The same question without scaffolding will not ask you intermediate questions but will instead just ask you to find the final answer. Part of the process of answering them involves understanding the question and planning the steps involved in finding the solution. However, not all questions are like this. Some are more traditional in that there is scaffolding, and it is fairly easy to see what needs to be done to answer them. In this book we will cover both types of questions.

Other maths revision books don't explore this at all and treat the answering of these questions in the same way as you would with the old specification.

Like most revision books, we will go through worked examples, but you will be encouraged to analyse the question before you start and be able to plan what you have to do.

As part of the worked examples, there will be tips which will enable you to see the way forward with answers. Also included will be 'watch outs' which will be points or misconceptions made by students.

Knowledge and understanding

Topics start with a list of material you need to know from your AS studies. You will need to understand all the AS material first before you start your A2 revision.

There is then a section called 'Quick revision' which summarises the material you need to know.

Looking at exam questions

Answering full examination questions lies at the heart of this book so rather than look at small parts of questions, we will instead be concentrating on complete examination questions.

Rather than go straight to the answers, we will instead look at the analysis of the question and the thought processes involved in deciding how to answer the question.

Each question is divided into the following parts:

- The question.
- Thinking about the question – what does the question involve? What specification topics does it involve?
- Starting the solution – planning the steps you need to take to answer the questions.
- The solution.

The examination questions are also annotated with advice about the knowledge, skills and techniques needed to answer them.

Exam practice questions

These questions are of examination standard and they allow you to have a go at them on your own. Full solutions are provided, showing you the steps to take.

Planning your revision and effective ways to revise

Revision is very much a personal thing – some people revise more effectively listening to music, while others need complete silence. Here are a few tips that may work for you:

Check you have completed all the material in each topic – check you have completed all the topics in the specification by printing out a copy of the specification from the WJEC website and match the material against your notes.

Ensure sure you understand the basics – fractional indices, the solution of simultaneous and quadratic equations, completing the square, etc. Remember that your knowledge of AS Maths will be assumed.

Target your revision – have a firm understanding of what you know and evaluate your strengths and weaknesses. Don't spend too much time going through things you know, but instead concentrate on your weaknesses.

Maximise your time – try to concentrate on your revision without breaking off to check your email, social media, etc. You may find it easier to work in a library where you won't be distracted.

Use technology – if there is no teacher to ask about a solution to a question or to ask about a particular topic, use sites such as YouTube where there are videos on many different A2 Maths topics.

Use your time effectively – use your phone to check formulae you need to remember or for explanations of topics you are unsure of during any spare time you have.

The use of calculators

All the papers in A2 Maths allow you to use a calculator. You will be at a disadvantage if you do not use a more advanced calculator than the one you probably used for your GCSE studies. Using the advanced calculator, you will be able to compute summary statistics and access probabilities from standard statistical distributions. You can also easily solve simultaneous and quadratic equations using the calculator.

Which calculator is recommended? – you have to be careful that the one you buy is allowed by the examination board. Your teacher/ lecturer probably recommended one for you and the most popular one by far is the CASIO fx-991EX CLASSWIZ.

A bit of advice, though, don't purchase a new calculator just before the exam. You need practice at using it. Also, it might be worth buying a book which explains how to use the calculator in a way that is useful for A-Level Maths or look at some of the many videos on YouTube. You will find lots of useful shortcuts and ways of using the calculator that will enable you to save time as well as improve your accuracy when answering questions.

Information about your examination

There are two units, and both must be taken.

A2 Unit 3: Pure Mathematics B – a written examination 2 hours 30 minutes and worth 120 marks.

A2 Unit 4: Applied Mathematics B – a written examination 1 hour 45 minutes and worth 80 marks. This unit is divided into two sections:

Section A: Statistics (40 marks)

Section B: Mechanics (40 marks)

Calculators can be used in the exam for both the units.

It is up to you how you divide your time between sections A and B.

Formulae and identities you must remember

Some formulae and identities are included on the formula sheet. Here are the formulae and identities that are not included on the formulae sheet and therefore must be remembered.

Pure mathematics

Quadratic equations

$ax^2 + bx + c = 0$ has roots $\dfrac{-b \pm \sqrt{b^2 - 4ac}}{2a}$

Laws of indices

$$a^x \times a^y \equiv a^{x+y}$$
$$a^x \div a^y \equiv a^{x-y}$$
$$(a^x)^y \equiv a^{xy}$$

Laws of logarithms

$$x = a^n \iff n = \log_a x \text{ for } a > 0 \text{ and } x > 0$$
$$\log_a x + \log_a y \equiv \log_a xy$$
$$\log_a x - \log_a y \equiv \log_a\left(\frac{x}{y}\right)$$
$$k \log_a x \equiv \log_a x^k$$

Series

n^{th} term of an arithmetic series $\quad t_n = a + (n-1)d$

n^{th} term of a geometric series $\quad t_n = ar^{n-1}$

Coordinate geometry

A straight-line graph, gradient m passing through (x_1, y_1) has equation
$$y - y_1 = m(x - x_1)$$
Straight lines with gradients m_1 and m_2 are perpendicular when
$$m_1 m_2 = -1$$

Trigonometry

In the triangle ABC

Sine rule: $\quad \dfrac{a}{\sin A} = \dfrac{b}{\sin B} = \dfrac{c}{\sin C}$

Cosine rule: $\quad a^2 = b^2 + c^2 - 2bc \cos A$

$$\text{Area} = \tfrac{1}{2} ab \sin C$$

Why not take a picture of all these formulae and equations on your phone and keep referring to them. Hopefully, after a while you will remember them.

Trigonometric identities

$$\cos^2\theta + \sin^2\theta = 1$$

$$\tan\theta = \frac{\sin\theta}{\cos\theta}$$

$$\sec^2\theta = 1 + \tan^2\theta$$

$$\text{cosec}^2\theta = 1 + \cot^2\theta$$

sec, cosec and cot

$$\sec\theta = \frac{1}{\cos\theta}$$

$$\text{cosec}\,\theta = \frac{1}{\sin\theta}$$

$$\cot\theta = \frac{1}{\tan\theta}$$

Double angle formulae

$$\sin 2A = 2\sin A\cos A$$

$$\cos 2A = \cos^2 A - \sin^2 A$$

$$= 1 - 2\sin^2 A$$

$$= \cos^2 A - 1$$

$$\tan 2A = \frac{2\tan A}{1 - \tan^2 A}$$

Use of small angle approximation for sine, cosine and tangent

If the angle θ is small and measured in radians:

$$\sin\theta \approx \theta$$

$$\cos\theta \approx 1 - \frac{\theta^2}{2}$$

$$\tan\theta \approx \theta$$

Mensuration

Circumference and area of circle, radius r and diameter d:

$$C = 2\pi r = \pi d \qquad A = \pi r^2$$

Pythagoras' theorem: In any right-angled triangle where a, b and c are the lengths of the sides and c is the hypotenuse:

$$c^2 = a^2 + b^2$$

Area of trapezium = $\frac{1}{2}(a + b)h$, where a and b are the lengths of the parallel sides and h is their perpendicular separation.

Volume of a prism = area of cross section × length

For a circle of radius, r, where an angle at the centre of θ radians subtends an arc of length s and encloses an associated sector of area A:

$$s = r\theta \qquad A = \tfrac{1}{2}r^2\theta$$

Differentiation

Function	Derivative
x^n	nx^{n-1}
e^{kx}	ke^{kx}
a^{kx}	$ka\ln a$
$\sin^{kx}$	$k\cos kx$
$\cos^{kx}$	$-k\sin kx$
$\tan^{kx}$	$k\sec^2 kx$
$\ln x$	$\dfrac{1}{x}$

The Product rule

If, $\qquad y = f(x)\,g(x), \qquad \dfrac{dy}{dx} = f(x)\,g'(x) + g(x)\,f'(x)$

The Chain rule

If y is a function of u and u is a function of x, then the chain rule states:

$$\frac{dy}{dx} = \frac{dy}{du} \times \frac{du}{dx}$$

Integration

Function	Integral
x^n	$\dfrac{1}{n+1}x^{n+1} + c,\, n \neq 0$
e^{kx}	$\dfrac{e^{kx}}{k} + c$
$\dfrac{1}{x}$	$\ln\lvert x\rvert + c$
$\sin kx$	$-\dfrac{1}{k}\cos kx + c$
$\cos kx$	$\dfrac{1}{k}\sin kx + c$
$(ax+b)^n$	$\dfrac{(ax+b)^{n+1}}{(n+1)a} + c \quad (n \neq -1)$
e^{ax+b}	$\dfrac{e^{ax+b}}{a} + c$
$\dfrac{1}{ax+b}$	$\dfrac{1}{a}\ln\lvert ax+b\rvert + c$
$\sin(ax+b)$	$\dfrac{-\cos(ax+b)}{a} + c$
$\cos(ax+b)$	$\dfrac{\sin(ax+b)}{a} + c$

$$\text{Area under a curve} = \int_a^b y\,dx \quad (y \geq 0)$$

Mechanics

Forces and equilibrium

$$\text{Weight} = \text{mass} \times g$$

Newton's second law in the form: $F = ma$

Kinematics

The equations of motion

$$v = u + at$$
$$s = ut + \tfrac{1}{2}at^2$$
$$v^2 = u^2 + 2as$$
$$s = \tfrac{1}{2}(u + v)t$$

s = displacement/distance
u = initial velocity/speed
v = final velocity/speed
a = acceleration
t = time

If the acceleration is zero, $\text{speed} = \dfrac{\text{distance travelled}}{\text{time taken}}$

Motion under variable acceleration

For motion in a straight line with variable acceleration:

$$v = \frac{dr}{dt} \qquad\qquad a = \frac{dv}{dt} = \frac{d^2r}{dt^2}$$

$$r = \int v \, dt \qquad\qquad v = \int a \, dt$$

Finding the magnitude of a vector

For a vector in two-dimensions such as $\mathbf{v} = a\mathbf{i} + b\mathbf{j}$,
Magnitude of vector $= |\mathbf{v}| = \sqrt{a^2 + b^2}$

For a vector in three-dimensions such as $\mathbf{v} = a\mathbf{i} + b\mathbf{j} + c\mathbf{k}$

Magnitude of vector $= |\mathbf{v}| = \sqrt{a^2 + b^2 + c^2}$

Statistics

$$\bar{x} = \frac{\sum x}{n} = \frac{\sum fx}{\sum f}$$

$$z = \frac{\bar{x} - \mu}{\frac{\sigma}{\sqrt{n}}} \text{ for a sample}$$

Formulae included on the formula sheet

The following formulae need not be remembered as they are included on the formula booklet that will be given to you in the exam.

Remember that all the formulae needed for AS will also be needed for A2.

Pure Mathematics

Mensuration

Surface area of sphere $= 4\pi r^2$

Area of curved surface of cone $= \pi r \times$ slant height

Arithmetic series

$$S_n = \frac{1}{2}n(a + l) = \frac{1}{2}n\,[2a + (n - 1)d]$$

Geometric series

$$S_n = \frac{a\,(1 - r^n)}{1 - r}$$

$$S_\infty = \frac{a}{1 - r} \text{ for } |r| < 1$$

Binomial series

$$(a + b)^n = a^n + \binom{n}{1}a^{n-1}b + \binom{n}{2}a^{n-2}b^2 + \dots + \binom{n}{r}a^{n-r}b^r + \dots + b^n \quad (n \in \mathbb{N})$$

where $\binom{n}{r} = {}^nC_r = \dfrac{n!}{r!(n - r)!}$

$$(1 + x)^n = 1 + nx + \frac{n(n - 1)}{1 \times 2}x^2 + \dots + \frac{n(n - 1) \dots (n - r + 1)}{1 \times 2 \times \dots \times r}x^r \quad (|x| < 1, n \in \mathbb{N})$$

Trigonometric identities

$$\sin (A \pm B) = \sin A \cos B \pm \cos A \sin B$$

$$\cos (A \pm B) = \cos A \cos B \mp \sin A \sin B$$

$$\tan (A \pm B) = \frac{\tan A \pm \tan B}{1 \mp \tan A \tan B}$$

Differentiation

Function	Derivative
$\dfrac{f(x)}{g(x)}$	$\dfrac{f'(x)\,g(x) - f(x)\,g'(x)}{(g(x))^2}$
$\tan x$	$\sec^2 x$
$\sec x$	$\sec x \tan x$
$\cot x$	$-\cosec^2 x$
$\cosec x$	$-\cosec x \cot x$
$\sin^{-1} x$	$\dfrac{1}{\sqrt{1 - x^2}}$
$\cos^{-1} x$	$-\dfrac{1}{\sqrt{1 - x^2}}$
$\tan^{-1} x$	$\dfrac{1}{1 + x^2}$

Integration (+ constant where relevant)

Function	Integral
$\sec^2 x$	$\tan x$

$$\int u \frac{dv}{dx}\,dx = uv - \int v \frac{du}{dx}\,dx$$

Numerical integration

The Trapezium rule: $\int_a^b y\,dx \approx \frac{1}{2}h\,\{(y_0 + y_n) + 2\,(y_1 + y_2 + \ldots + y_{n-1})\}$,

where $h = \dfrac{b - a}{n}$

Numerical Solution of Equations

The Newton–Raphson iteration for solving $f(x) = 0$: $\quad x_{n+1} = x_n - \dfrac{f(x_n)}{f'(x_n)}$

Vectors

The point dividing AB in the ratio $\lambda : \mu$ is $\dfrac{\mu \mathbf{a} + \lambda \mathbf{b}}{\lambda + \mu}$

Mechanics

(You must memorise all the formulae needed for Mechanics)

Probability and statistics

Probability

$$P(A \cup B) = P(A) + P(B) - P(A \cap B)$$

Standard discrete distributions

Distribution of X	$P(X = x)$	Mean	Variance
Binomial B(n, p)	$\binom{n}{x} p^x (1 - p)^{n-x}$	np	$np(1 - p)$
Poisson Po(λ)	$e^{-\lambda} \dfrac{\lambda^x}{x!}$	λ	λ

Sampling distributions

For a random sample $X_1, X_2, \ldots, X_n$ of n independent observations from a distribution having mean μ and variance σ^2

$\overline{X}$ is an unbiased estimator of μ, with $\mathrm{Var}\left(\overline{X}\right) = \dfrac{\sigma^2}{n}$

S^2 is an unbiased estimator of σ^2, where $S^2 = \dfrac{\sum\left(X_i - \overline{X}\right)^2}{n - 1}$

Standard continous distributions

Distribution of X	P.D.F	Mean	Variance
Uniform (Rectangular) on $[a, b]$ U$[a, b]$	$\dfrac{1}{b - a}$	$\dfrac{1}{2}(a + b)$	$\dfrac{1}{12}(b - a)^2$
Normal N(μ, σ^2)	$\dfrac{1}{\sigma\sqrt{2\pi}} e^{-\frac{1}{2}\left(\frac{x - \mu}{\sigma}\right)^2}$	μ	σ^2
Exponential Exp (λ)	$\lambda e^{-\lambda x}$	$\dfrac{1}{\lambda}$	$\dfrac{1}{\lambda^2}$

Tips for the actual exam

Do not change your calculator to a different model just before your exam. Make sure you understand how to use all the functions, especially the statistical functions as the exam itself is not the place to start learning them.

Pace yourself and avoid getting bogged down on a question. Note that you do not have to do questions in the order they are presented on the exam paper.

Include answer steps with clear explanations.

If you feel confident – include shortcuts to workings out – however, remember that time saved sometimes comes at the expense of accuracy.

Tidy up final answers – remember to cancel fractions, simplify algebraic expressions, give answers to an appropriate accuracy, etc.

Draw clearly labelled diagrams, if appropriate. Always draw graphs for coordinate geometry questions even if they are not asked for.

Watch out when drawing curves with asymptotes that the curves do not touch the asymptotes.

In trigonometric proofs, do not work with both sides together. It is best to start with the more complicated side on its own and then prove that it is equal or equivalent to the less complicated side.

For example, to prove:

$$\frac{\sin^3 \theta + \sin \theta \cos^2 \theta}{\cos \theta} \equiv \tan \theta$$

you would start with the more complicated side (i.e. the left-hand side) and manipulate it to prove the less complicated right-hand side.

Make it clear what your answer is – the examiner should not have to search for your answer amongst a load of working out.

Good luck with your revision and for the exam itself.

Steve Doyle

1 Proof

Prior knowledge

You will need to make sure you understand the following from your AS studies:

- The meanings of real, imaginary, rational and irrational numbers.
- The solutions of quadratic inequalities.

Quick revision

Proof by contradiction

Here you are given a conjecture which you then assume is false. You then use this assumption and show that it leads to a contradiction.

This is best explained by this example:

Prove by contradiction the following proposition:

When x and y are integers, $8x + 2y \neq 5$

This is the conjecture.

Assuming there are integer values of x and y for which $8x + 2y = 5$.

Here we assume the conjecture is false.

$$8x + 2y = 5$$
$$2(4x + y) = 5$$
$$4x + y = \frac{5}{2}$$

We now take steps using algebraic manipulation to produce a contradiction which means that the assumption is false.

Now for all integer values of x and y, $4x + y$ would be an integer so not a fraction.

As this is a contradiction, the assumption is incorrect, so $8x + 2y \neq 5$.

Hints and tips when using proof by contradiction

Here are a few things that crop up in proof by contradiction that you need to know:

If a is an odd non-zero integer, then a^2 is an odd non-zero integer.

If a is an even non-zero integer, then a^2 is an even non-zero integer.

If x is an even non-zero integer, then it can be written as $x = 2k$ where k is a non-zero integer.

If x is an odd non-zero integer, then it can be written as $x = 2k + 1$ where k is a non-zero integer.

Looking at exam questions

1   Prove by contradiction the following proposition. When x is real and positive,

$$x + \frac{49}{x} \geq 14$$

The first line of the proof is given below.

Assume that there is a positive and real value of x such that:

$$x + \frac{49}{x} < 14$$

[4]

(WJEC June 2008 C4 q10)

Thinking about the question

The assumption is given in the question and we have to manipulate the inequality until we obtain a contradiction. We need to manipulate the given equality to obtain an inequality that is in a form that could be solved.

Starting the solution

If both sides of the inequality are multiplied by x we can remove the x in the denominator. This will result in a quadratic inequality which will be in the form $ax^2 + bx + c < 0$.

We can then factorise the quadratic and look for a contradiction in the result.

The solution

$$x + \frac{49}{x} < 14$$

Multiplying both sides by x, we obtain:

$$x^2 + 49 < 14x$$

$$x^2 - 14x + 49 < 0$$

$$(x - 7)^2 < 0$$

For $(x - 7)^2$ to be less than zero, $x - 7$ would have to be an imaginary number. This contradicts that $x - 7$ and x are real.

Hence $\qquad x + \frac{49}{x} \geq 14$

> It should be possible to form a quadratic inequality by manipulating this inequality.

> Notice that the assumption is given in the question. You just have to carry on the steps that will lead you to a contradiction.

---

2 Use proof by contradiction to prove that:

$1 + \sqrt{3}$ is irrational. [4]

Thinking about the question

We first need to consider the difference between a rational and an irrational number. Rational numbers can be expressed as fractions (i.e. $\frac{a}{b}$).

Starting the solution

We start with the assumption that $1 + \sqrt{3}$ is rational and therefore can be expressed as the fraction $\frac{a}{b}$ and then equate these two. By manipulating this equation, we should be able to obtain a contradiction.

Here we assume the conjecture in the question is false.

Here we manipulate this equation so that we have an irrational part on the left. If we can prove that the right is rational, we have a contradiction.

The solution

Assuming that $1 + \sqrt{3}$ is rational.

$1 + \sqrt{3}$ can be expressed as the fraction $\frac{a}{b}$, where a and b are integer rational numbers greater than 0.

$$1 + \sqrt{3} = \frac{a}{b}$$

$$\sqrt{3} = \frac{a}{b} - 1$$

$$\sqrt{3} = \frac{a - b}{b}$$

Now, as $a - b$ will be a rational integer and b is a rational integer, it means $\frac{a - b}{b}$ will be a rational number.

This is a contradiction as $\sqrt{3}$, which is equal to $\frac{a - b}{b}$, is irrational.

Hence, $1 + \sqrt{3}$ is irrational.

Exam practice

1. Use proof by contradiction to prove that for all integer values of x and y, there are no values for which:
$$9x + 12y = 1$$ [2]

2. Complete the following proof by contradiction to show that, if n is a positive integer and $3n + 2n^3$ is odd, then n is odd. [2]
It is given that $3n + 2n^3$ is odd.
Assume that n is even so that n = 2k.
(WJEC June 2007 C4 q10)

3. Prove by contradiction that if $2n^2 + n$ is even, then n is even. [3]

4. Prove by contradiction that if x and y are both real then:
$$x^2 + 4y^2 \geq 4xy.$$ [3]

5. Use proof by contradiction to prove that $\sqrt{3}$ is irrational. [4]

6. Complete the following proof by contradiction to show that $x + \frac{25}{x} \geq 10$ when x is real and positive.
Assume that $x + \frac{25}{x} < 10$ when x is real and positive.
Since x is positive, multiplication of both sides of the inequality by x gives:
$$x^2 + 25 < 10x.$$ [4]
(WJEC June 2005 C4 Q10)

7 Prove by contradiction the following proposition:

   If $(x - 3)(x + 1) < 0$ then $-1 < x < 3$.

   The first line of the proof is given below.

   Assume that if $(x - 3)(x + 1) \geq 0$ then $x \leq -1$ or $x \geq 3$.   [4]

8 Prove by contradiction the following proposition. When x and y are real positive integers,

   $y^2 - 4x - 3 \neq 0$.   [4]

2 Algebra and functions

Prior knowledge

You will need to make sure you fully understand the following from your AS studies:

- Sketching the graph of a quadratic function.

- Solving linear and quadratic inequalities.

- Sketching curves of functions.

- Transformations of the graph of $y = f(x)$.

- Exponential and logarithmic graphs.

Quick revision

Partial fractions

$$\frac{\alpha x + \beta}{(cx + d)(ex + f)} \equiv \frac{A}{cx + d} + \frac{B}{ex + f} \qquad \text{(i)}$$

$$\frac{\alpha x^2 + \beta x + \gamma}{(cx + d)(ex + f)^2} \equiv \frac{A}{cx + d} + \frac{B}{ex + f} + \frac{C}{(ex + f)^2} \qquad \text{(ii)}$$

In both cases, clear the fractions and choose appropriate values of x.

In (ii), an equation involving coefficients of x^2 may be used.

Functions

A function is a relation between a set of inputs and a set of outputs such that each input is related to exactly one output.

The domain and range of a function

The domain is the set of input values that can be entered into a function.

The range is the set of output values from a function.

Composition of functions

Composition of functions involves applying two or more functions in succession.

The composite function $fg(x)$ means $f(g(x))$ and is the result of performing the function g first and then f.

One-to-one functions

A function where one output value would correspond to only one possible input value.

To find $f^{-1}(x)$ given $f(x)$

First check that $f(x)$ is a one-to-one function. Let y equal the function and rearrange so that x is the subject of the equation. Replace x on the left with $f^{-1}(x)$ and on the right replace all occurrences of y with x.

Domain and range of inverse functions

The range of $f^{-1}(x)$ is the same as the domain of $f(x)$.

The domain of $f^{-1}(x)$ is the same as the range of $f(x)$.

Graphs of inverse functions

The graph of $y = f^{-1}(x)$ is obtained by reflecting the graph of $y = f(x)$ in the line $y = x$.

The modulus function

The modulus of x is written $|x|$ and means the numerical value of x (ignoring the sign).

So whether x is positive or negative, $|x|$ is always positive (or zero).

Graphs of modulus functions

First plot the graph of $y = f(x)$ and reflect any part of the graph below the x-axis in the x-axis. The resulting graph will be $y = |f(x)|$.

Combinations of transformations

If a graph of $y = f(x)$ is drawn, then the graph of $y = f(x - a) + b$ can be obtained by applying the translation $\binom{a}{b}$ to the original graph.

If a graph of $y = f(x)$ is drawn, then the graph of $y = af(x - b)$ can be obtained by applying the following two transformations in either order: a stretch parallel to the y-axis with scale factor a and a translation of $\binom{b}{0}$.

If a graph of $y = f(x)$ is drawn, then the graph of $y = f(ax)$ can be obtained by scaling the x values by $\frac{1}{a}$.

The function e^x and its graph

$y = e^x$

e^x is a one-to-one function and has an inverse $\ln x$.

The graph of $y = e^x$ cuts the y-axis at $y = 1$ and has the x-axis as an asymptote.

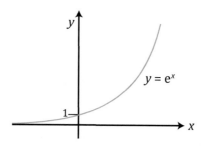

$D(f) = (-\infty, \infty)$

$R(f) = (0, \infty)$

The function ln x and its graph

$y = \ln x$

> $\ln x$ is a one-to-one function and has an inverse e^x.
>
> The graph of $y = \ln x$ cuts the x-axis at $x = 1$ and has the y-axis as an asymptote.

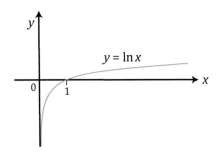

> $D(f) = (0, \infty)$
>
> $R(f) = (-\infty, \infty)$

The functions $y = \ln x$ and $y = e^x$ are inverse functions.

So, the graphs of $y = \ln x$ and $y = e^x$ are reflections of each other in the line $y = x$.

Also, $e^{\ln x} = x$ and $\ln e^x = x$.

Looking at exam questions

1   Solve the following

   (a) $|3x - 8| \leq +4$ [3]

   (b) $\sqrt{3|x| - 5} = 10$ [2]

Thinking about the question

Both of these questions involve the modulus function. We have to remember that this will usually result in two inequalities for x. We need to check that these inequalities could be combined into a single inequality.

Starting the solution

For part (a) we notice that the modulus sign applies to the entire expression on the left-hand side. We first need to remove the modulus sign and then find the two values for x.

For part (b) we need to remove the square root on the left-hand side by squaring both sides. We can then solve the resulting equation for x.

2 Algebra and functions

The solution

(a) $|3x - 8| \leq -4$

$3x - 8 \leq -4$ and $3x - 8 \geq 4$

$3x \leq 4$ and $3x \geq 12$

$x \leq \frac{4}{3}$ and $x \geq 4$

(b) $\sqrt{3|x| - 5} = 10$

Squaring both sides gives

$3|x| - 5 = 100$

$3|x| = 105$

$|x| = 35$

$x = \pm 35$

---

2 (a) Find the range of values of x for which $|1 - 3x| > 7$. [3]

(b) Sketch the graph of $y = |1 - 3x| - 7$. Clearly label the minimum point and the points where the graph crosses the x-axis. [4]

(WJEC June 2019 Unit 3 q5)

Thinking about the question

Part (a) is a standard question where you solve an inequality involving a modulus. There are two methods to do this. One method involves first squaring both sides and solving the resulting quadratic inequality. The other involves removing the inequality and adding a ± to the number on the right of the inequality.

Part (b) involves a modulus graph.

Starting the solution

For part (a) we will use the method that involves squaring both sides. We then need to obtain a quadratic inequality which can then be solved. It is noted that the curve will be U-shaped and because of the > sign we need the regions lying above the x-axis.

For part (b) we need to first consider the graph of $y = 1 - 3x$ and find where the line cuts both axes. The part of the line lying below the x-axis is reflected in the x-axis to give $|1 - 3x|$. The -7 in $y = |1 - 3x| - 7$ means we need to apply a translation of $\begin{pmatrix} 0 \\ -7 \end{pmatrix}$ to the graph which shifts the graph 7 units vertically down.

The solution

(a) $|1 - 3x| > 7$

Squaring both sides, we obtain

$$(1 - 3x)^2 > 49$$

$$1 - 6x + 9x^2 > 49$$

$$9x^2 - 6x - 48 > 0$$

$$3x^2 - 2x - 16 > 0$$

$$(3x - 8)(x + 2) > 0$$

Hence $x < -2$ or $x > \frac{8}{3}$

(b) $y = |1 - 3x| - 7$

We first consider the line $y = 1 - 3x$

When $x = 0, y = 1$ and when $y = 0, x = \frac{1}{3}$

We then draw the line as shown but reflecting any part that lies below the x-axis in the x-axis.

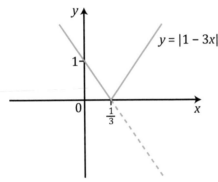

To take account of the –7 we translate the graph by $\begin{pmatrix} 0 \\ -7 \end{pmatrix}$ which moves it 7 units downwards.

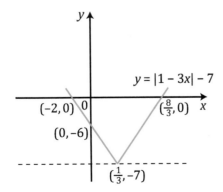

To find the points of intersection with the axes we can use the two equations $y = -(1 - 3x) - 7$ and $y = 1 - 3x - 7$ and equate both to zero to find the points of intersection with the x-axis.

3   On the same set of axes sketch the graphs of $y = 2|x|$ and $y = |x - 3|$.

Hence, solve the inequality $2|x| \leq |x - 3|$.   [5]

Thinking about the question

Both of the given equations are those of straight lines, so we need to find some points of intersection with the axes and then we need to consider how to apply the modulus sign to each graph.

Starting the solution

To draw the graphs, draw the section below the x-axis with a dotted line. We can then reflect the section of the graph below the x-axis in the x-axis to complete the lines for the two graphs. We then need to find the range of values where the graph for $y = 2|x|$ is below or on the graph for $y = |x - 3|$.

The solution

$y = 2x$ is a straight line through the origin with gradient $= 2$.
$y = x - 3$ is a line that intercepts the y-axis at -3 and the x-axis at 3. The modulus of each of these lines will reflect those parts of the lines lying below the x-axis in the x-axis.

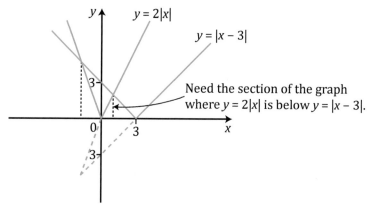

Need the section of the graph where $y = 2|x|$ is below $y = |x - 3|$.

We need to find the x-values where the two lines intersect. We can do this by equating the two equations and solving them for x.

$$2|x| = |x - 3|$$

> On squaring, the modulus signs can be removed.

Squaring both sides, we obtain:

$$4x^2 = x^2 - 6x + 9$$

$$3x^2 + 6x - 9 = 0$$

$$x^2 + 2x - 3 = 0$$

$$(x + 3)(x - 1) = 0$$

$$x = 1 \text{ or } -3$$

> A graph of $y = x^2 + 2x - 3$ is U-shaped and cuts the x-axis at $x = 1$ and -3. As we need to find $y \leq x^2 + 2x - 3$ we need the region of the curve below the x-axis. The region is between, and including, $x = 1$ and -3.

From the graph, the range is $-3 \leq x \leq 1$

---

4 The function f has domain $(-\infty, 12]$ and is defined by:

$$f(x) = e^{4-\frac{x}{3}} + 8$$

(a) Find an expression for $f^{-1}(x)$. [4]

(b) Write down the domain of f^{-1}. [2]

(WJEC June 2016 C3 q9)

Thinking about the question

In the domain $(-\infty, 12]$ the $-\infty$ part means the lowest value approaches $-\infty$ and the 12 means the highest value equals 12.

Starting the solution

For part (a) we need to follow the steps for finding the inverse of a function. As part of this involves finding x we notice that x is included in the power of e which means we will use ln to obtain x.

The first step in finding the inverse is to let y equal the function.

The solution

(a) Let $\quad y = e^{4-\frac{x}{3}} + 8$

$$y - 8 = e^{4-\frac{x}{3}}$$

$$\ln(y-8) = 4 - \frac{x}{3}$$

$$\frac{x}{3} = 4 - \ln(y-8)$$

$$x = 3[4 - \ln(y-8)]$$

$$f^{-1}(x) = 3[4 - \ln(x-8)]$$

In the following steps we rearrange the equation so that x is the subject.

We now replace the x with $f^{-1}(x)$.

(b) $D(f^{-1}) = [9,\infty)$

---

5 (a) Given that f is a function,

(i) State the condition for f^{-1} to exist,

(ii) Find $ff^{-1}(x)$. [2]

(b) The functions g and h, are given by:

$$g(x) = x^2 - 1,$$

$$h(x) = e^x + 1.$$

(i) Suggest a domain for g such that g^{-1} exists.

(ii) Given the domain of h is $(-\infty, \infty)$, find an expression for $h^{-1}(x)$ and sketch, using the same axes, the graphs of $h(x)$ and $h^{-1}(x)$. Indicate clearly the asymptotes and the points where the graphs cross the coordinate axes.

(iii) Determine an expression for $gh(x)$ in its simplest form. [8]

(WJEC June 2018 Unit 3 q12)

The domain of f^{-1} equals the range of f so to find the range of f the domain values are substituted for x in the function. When x approaches $-\infty$, $e^{4-\frac{x}{3}}$ approaches ∞ and adding 8 to it still makes it approach ∞. When $x = 12$, $f(12) = e^{4-\frac{x}{3}} + 8 = e^0 + 8 = 9$. So the range of f is $[9,\infty)$ which is the same as the domain of f^{-1}.

Thinking about the question

This question covers most of the theory of functions, so we will need to recall information about the condition for a function to have an inverse and what happens when we apply an inverse and a function in succession. Part (b) involves drawing graphs so we need to understand the shapes of the graphs from their equations and also recall the way transformations can be applied to these graphs.

Starting the solution

For part (a) we need to recall some factual information about functions. For the inverse of a function to exist, the function must be a one-to-one function. When we apply an inverse function and then the function they cancel out and leave x unchanged. For example, if the original function was squaring x, the inverse function would be square rooting. Applying these in either order would result in x being unchanged.

For part (b) (i) we need to apply the statement made in part (a)(i).

For part (b) (ii) we note the shape of the graph which is an e^x curve shifted vertically upwards by 1. It is therefore a one-to-one function. We need to recall that to sketch the inverse we will need to reflect the curve in the line $y = x$. We will need to find $h^{-1}(x)$ and then use this to determine the point of intersection with the axis and also the asymptote.

Part (b)(iii) involves substituting $e^x + 1$ for x in the function $x^2 - 1$.

The solution

(a) (i)   The function f must be a one-to-one function.

   (ii) $ff^{-1}(x) = x$

(b) (i)   Sketching the graph of $g(x)$, we obtain:

This means a value for $f(x)$ would correspond only to a single value of x.

Applying an inverse and then the original function (or vice versa) results in x remaining unchanged.

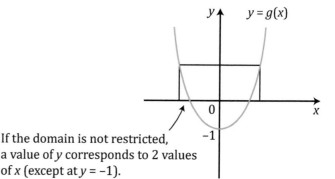

If the domain is not restricted, a value of y corresponds to 2 values of x (except at $y = -1$).

We need to restrict the domain to either side of the y-axis.

For $g^{-1}(x)$ to exist, $g(x)$ must be a one-to-one function. So, one value of $g(x)$ must correspond to only a single value of x.

Hence $g^{-1}(x)$ exists if the domain is $[0, \infty)$ or $(-\infty, 0]$.

(ii) Let $\quad y = e^x + 1$
$$y - 1 = e^x$$
$$\ln(y - 1) = x$$
$$h^{-1}(x) = \ln(x - 1)$$

These are the steps used to find the inverse of a function.

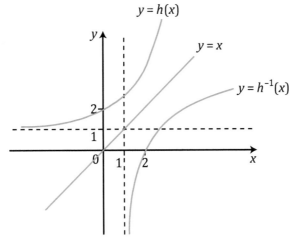

(iii) $gh(x) = (e^x + 1)^2 - 1$
$$= e^{2x} + 2e^x + 1 - 1$$
$$= e^{2x} + 2e^x$$

Exam practice

1 The diagram below shows a sketch of the graph of $y = f(x)$. The graph passes through the points $(-2, 0)$, $(0, 8)$, $(4, 0)$ and has a maximum point at $(1, 9)$.

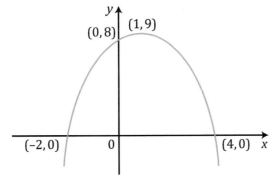

(a) Sketch the graph of $y = 2f(x + 3)$. Indicate the coordinates of the stationary point and the points where the graph crosses the x-axis. [3]

(b) Sketch the graph of $y = 5 - f(x)$. Indicate the coordinates of the stationary point and the point where the graph crosses the y-axis. [3]

(WJEC June 2018 Unit 3 q3)

2   Solve:
$$5|x + 1| - 10 = 15$$ [3]

3   Solve the following:
$$\frac{2|x| + 1}{1 - |x|} = 5$$ [4]

4   Solve the following:
(a) $|4x - 3| \geq 7$ [2]
(b) $\sqrt{2|x| + 2} = 4$ [3]

5   The diagram below shows a sketch of the graph of $y = f(x)$. The graph crosses the y-axis at the point $(0, -2)$, and the x-axis at the point $(8, 0)$.

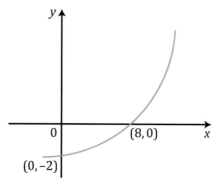

(a) Sketch the graph of $y = -4f(x + 3)$. Indicate the coordinates of the point where the graph crosses the x-axis and the y-coordinate of the point where $x = -3$.   [3]
(b) Sketch the graph of $y = 3 + f(2x)$. Indicate the y-coordinate of the point where $x = 4$.   [2]

(WJEC June 2019 Unit 3 q7)

6   The function $f(x)$ is defined by:
$$f(x) = \frac{\sqrt{x^2 - 1}}{x}$$
with domain $x \geq 1$.
(a) Find an expression for $f^{-1}(x)$. State the domain for f^{-1} and sketch both $f(x)$ and $f^{-1}(x)$ on the same diagram.   [6]
(b) Explain why the function $ff(x)$ cannot be formed.   [1]

(WJEC June 2019 Unit 3 q11)

7   Functions f and g are defined as follows:

$f(x) = \sqrt{x + 5}$

$g(x) = 3x^2 - 2$

If $D(f) = [-4, \infty)$ and $D(g) = [0, \infty)$

(a) Write down the range of f and the range of g.   [4]

(b) Find an expression for $gf(x)$ giving your answer in
    simplified form.   [2]

(c) Solve the equation $fg(x) = 9$.   [3]

8   The function f has domain $x \geq 1$ and is defined by:

$$f(x) = x - \frac{1}{x}$$

(a) Show that $f'(x)$ is always positive. Deduce the least value
    of $f(x)$.   [3]

(b) Find the range of f.   [1]

(c) The function g has domain $[0, \infty)$ and is defined by
    $g(x) = 3x^2 + 2$.

    Solve the equation

    $$gf(x) = \frac{3}{x^2} + 8.$$ [4]

(WJEC C3 June 2006 q8)

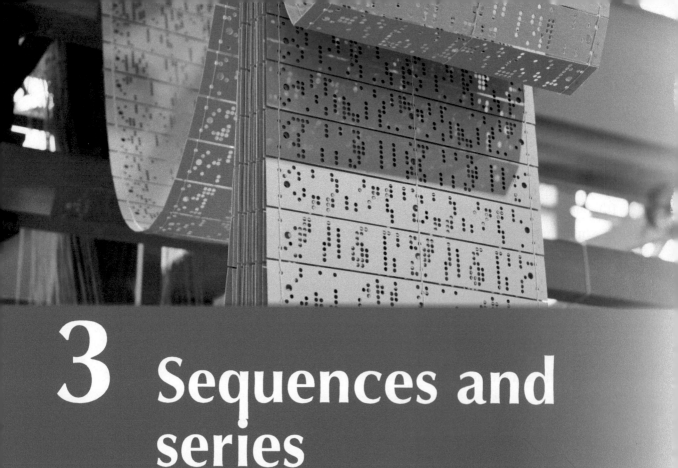

3 Sequences and series

Prior knowledge

You will need to understand the following from your AS studies:

- The binomial theorem.
- Use of the laws of logarithms.

Quick revision

The binomial expansion of $(a + b)^n$ for any rational value of n

$$(a + b)^n = a^n + \binom{n}{1}a^{n-1}b + \binom{n}{2}a^{n-2}b^2 + \dots + \binom{n}{r}a^{n-r}b^r + \dots + b^n$$

$$\binom{n}{r} = {}^nC_r = \frac{n!}{r!(n-r)!}$$

The binomial expansion of $(1 + x)^n$ for negative or fractional n

$$(1 + x)^n = 1 + nx + \frac{n(n-1)}{2!}x^2 + \frac{n(n-1)(n-2)}{3!}x^3 + \dots \qquad |x| < 1$$

Sequences, arithmetic series and geometric series

The nth term of an arithmetic sequence

The nth term $t_n = a + (n-1)d$

where a is the first term, d is the common difference and n is the number of terms.

The sum to n terms of an arithmetic series

$$S_n = \frac{n}{2}\big[2a + (n-1)d\big]$$

The nth term of a geometric sequence

The nth term $t_n = ar^{n-1}$

where a is the first term, r is the common ratio and n is the number of terms.

The sum to n terms of a geometric sequence

$$S_n = \frac{a(1 - r^n)}{1 - r} \quad \text{provided } r \neq 1$$

The sum to infinity of a geometric sequence

$$S_\infty = \frac{a}{1 - r}$$

Note that for the sum to infinity to exist $|r| < 1$

Looking at exam questions

1 The sixth term of a geometric sequence is 2187 and the fourth term is 243.

If the common ratio is positive, find the common ratio and the first term. [4]

Thinking about the question

This question is about the terms of a geometric sequence, so we need to recall the formula to give the nth term $t_n = ar^{n-1}$. This formula is not included in the formula booklet.

Starting the solution

We need to use the formula for the nth term and create two equations with a and r as the unknowns and solve these simultaneously to find r.

3 Sequences and series

Notice that by dividing these two simultaneous equations we can eliminate a.

Look at the question carefully to see if it is possible to have both values. The question tells us that r is positive.

The solution

$t_6 = ar^5 = 2187$

$t_4 = ar^3 = 243$

$\dfrac{t_6}{t_4} = \dfrac{ar^5}{ar^3} = r^2 = \dfrac{2187}{243} = 9$

$r = \pm 3$ but r is positive

Hence $r = 3$

---

2 A geometric series has first term a and common ratio r. The sum of the second and third terms of the series is −216. The sum of the fifth and sixth terms of the series is 8.

(a) Prove that $r = -\dfrac{1}{3}$. [5]

(b) Find the sum to infinity of the series. [3]

Thinking about the question

This question is about the terms of a geometric progression and we need to recall that the nth term $= ar^{n-1}$. This formula is not included in the formula booklet.

In part (b) we need to obtain the formula for the sum to infinity for a GP from the formula booklet.

Starting the solution

For part (a) we need to form a pair of equations in a and r and then solve them simultaneously to find the value of r.

For part (b) we need to substitute the value of r into one of the equations and then solve it to find the value of a. These two values are then substituted into the equation for the sum to infinity for a GP.

We now have a pair of equations that can be solved simultaneously to find the value of r. Note that you are not asked to find the value of a at this stage.

The solution

(a) $ar + ar^2 = -216$ (1)

$ar^4 + ar^5 = 8$ (2)

Multiplying equation (1) by r^3 we obtain:

$ar^4 + ar^5 = -216r^3$ (3)

Subtracting equation (3) from equation (2) we obtain:

$0 = 8 + 216r^3$

$r^3 = -\dfrac{8}{216}$

$r = -\dfrac{2}{6} = -\dfrac{1}{3}$

(b) $S_\infty = \dfrac{a}{1-r}$

Need to find the value of a.

$ar + ar^2 = -216$

$a(r + r^2) = -216$

$a\left(-\dfrac{1}{3} + \dfrac{1}{9}\right) = -216$

$a\left(-\dfrac{3}{9} + \dfrac{1}{9}\right) = -216$

$a\left(-\dfrac{2}{9}\right) = -216$

$a = 972$

$S_\infty = \dfrac{a}{1-r} = \dfrac{972}{1 - -\dfrac{1}{3}} = 729$

> The sum to infinity formula is included in the formula booklet.

---

3 (a) The 3rd, 19th and 67th terms of an arithmetic sequence form a geometric sequence.
Given that the arithmetic sequence is increasing and that the first term is 3, find the common difference of the arithmetic sequence. [5]

(b) A firm has 100 employees on a particular Monday. The next day it adds 12 employees onto its staff and continues to do so on every successive working day, from Monday to Friday.

(i) Find the number of employees at the end of the 8th week.

(ii) Each employee is paid £55 per working day. Determine the total wage bill for the 8-week period. [6]

(WJEC Unit 3 June 2019 q8)

Thinking about the question

For part (a) we need to obtain the formula for the nth term of an arithmetic progression. We also need the formula for the nth term of a geometric progression. This might involve using the three terms of the arithmetic progression to find expressions for the common ratio and then equate the common ratios.

For part (b) we can see that we are dealing with an arithmetic progression having first term, $a = 100$ and common difference, $d = 12$. We need to be careful to use a working week as 5 days.

Starting the solution

For part (a) we write equations for the 3rd, 19th and 67th terms of an arithmetic sequence and include the value of a (i.e. 3) in these equations. We can now divide the second term by the first term to find the common ratio of the geometric progression. We can also divide the third term by the second term to also find the common ratio. We can now equate these two expressions.

For part (b)(i) we need to form the terms of an arithmetic sequence and then find a and d and substitute these into the expression for the nth term in order to find the 8th term. For part (b)(ii) we use the first term and the common difference to find the sum to n terms of the arithmetic series. The formula is obtained from the formula booklet.

The solution

(a) For the arithmetic series:

3rd term = $3 + 2d$

19th term = $3 + 18d$

67th term = $3 + 66d$

These terms form a geometric sequence

$3 + 2d, 3 + 18d, 3 + 66d$

Common ratio $= \dfrac{3 + 18d}{3 + 2d} = \dfrac{3 + 66d}{3 + 18d}$

Now $\dfrac{3 + 18d}{3 + 2d} = \dfrac{1 + 22d}{1 + 6d}$

$(3 + 18d)(1 + 6d) = (1 + 22d)(3 + 2d)$

$3 + 18d + 18d + 108d^2 = 3 + 68d + 44d^2$

$64d^2 - 32d = 0$

$d(64d - 32) = 0$

$d = \dfrac{1}{2}$

(b) (i) $a = 100$ and $d = 12$

Sequence is 100, 112, 124, 136, 148, ...

End of 8th week is the 40th day

$t_{40} = a + (n - 1)d$

$\quad = 100 + (39)(12)$

$\quad = 568$

(ii) Wages will form an arithmetic series:

$100 \times 55 + 112 \times 55 + 124 \times 55 + ...$

$5500 + 6160 + 6820 + ...$

This forms an arithmetic series with first term, $a = 5500$ and common difference, $d = 660$.

$S_n = \dfrac{n}{2}\left[2a + (n - 1)d\right]$

$S_{40} = \dfrac{40}{2}\left[2(5500) + (39)660\right]$

$\quad = £734800$

> You can use the formula $t_n = a + (n - 1)d$, which needs to be remembered.

> Note that the other solution, $d = 0$, is ignored as this would make all the terms the same.

> The formula for S_n for an arithmetic series is included in the formula booklet.

4   The nth term of a number sequence is denoted by x_n.
The $(n + 1)$th term is defined by $x_{n+1} = 4x_n - 3$ and $x_3 = 113$.

(a)   Find the values of x_2 and x_1.   [2]

(b)   Determine whether the sequence is an arithmetic sequence, a geometric sequence or neither. Give reasons for your answer.
                                                                            [2]

(WJEC Unit 3 June 2019 q3)

Thinking about the question

This question concerns the generation of a sequence using a recurrence relation. To obtain the next term we put the value of the previous term into the recurrence relation (i.e. $4x_n - 3$).

For part (b) we need to establish if there is a common difference or common ratio between the terms of the sequence to decide whether it is an arithmetic or geometric sequence.

Starting the solution

For part (a) we use the value of x_3 (i.e. 113) to obtain the value of x_2 using the recurrence relation. Then, we use the value of x_2 with the recurrence relation to find the value of x_1.

For part (b) we can use the values of x_1, x_2 and x_3 to see if there is a common difference between consecutive terms to see whether or not it is an arithmetic or geometric sequence.

The solution

(a) $x_3 = 4x_2 - 3$ so $113 = 4x_2 - 3$ giving $x_2 = 29$

$x_2 = 4x_1 - 3$ so $29 = 4x_1 - 3$ giving $x_1 = 8$

(b) $x_3 - x_2 \neq x_2 - x_1$ so there is no common difference between the terms, so it is not an arithmetic sequence.

$\dfrac{x_3}{x_2} = \dfrac{113}{29}$

$\dfrac{x_2}{x_1} = \dfrac{29}{8}$

Now $\dfrac{113}{29} \neq \dfrac{29}{8}$ so there is no common ratio, so it is not a geometric sequence.

Hence it is not an arithmetic sequence or a geometric sequence.

5   Expand $\dfrac{4 - x}{\sqrt{1 + 2x}}$ in ascending powers of x up to and including the term in x^3. State the range of values of x for which the expansion is valid.   [6]

(WJEC Unit 3 June 2019 q2)

Thinking about the question

This is a question about binomial expansion and the formula for the expansion of $(1 + x)^n$ is obtained from the formula booklet.

Starting the solution

We need to bring the $\sqrt{1 + 2x}$ to the top and express it using an index. So $\sqrt{1 + 2x}$ in the denominator becomes $(1 + 2x)^{-\frac{1}{2}}$ in the numerator. This now multiplies the $(4 - x)$ to give $(4 - x)(1 + 2x)^{-\frac{1}{2}}$. We can then expand the $(1 + 2x)^{-\frac{1}{2}}$ using the binomial expansion formula. We just go as far as the term in x^3. We can then multiply $(4 - x)$ by the result of the expansion and do not include any terms higher than terms in x^3.

To find the range of values for which the expansion is valid, we recall that the expansion of $(1 + x)^n$ is valid for $|x| < 1$. In this case we need to replace x with $2x$ and then solve the resulting inequality.

The solution

The formula for the expansion of $(1 + x)^n$ is obtained from the formula booklet.

$$(1 + x)^n = 1 + nx + \frac{n(n - 1)x^2}{2!} + \frac{n(n - 1)(n - 2)x^3}{3!} + \ldots$$

$$\frac{4 - x}{\sqrt{1 + 2x}} = (4 - x)(1 + 2x)^{-\frac{1}{2}}$$

$$= (4 - x)[1 + (-\tfrac{1}{2})2x + \frac{(-\frac{1}{2})(-\frac{3}{2})(2x)^2}{2 \times 1} +$$

$$\frac{(-\frac{1}{2})(-\frac{3}{2})(-\frac{5}{2})(2x)^3}{3 \times 2 \times 1} + \ldots]$$

$$= (4 - x)[1 - x + \tfrac{3}{2}x^2 - \tfrac{5}{2}x^3]$$

$$= 4 - 4x + 6x^2 - 10x^3 - x + x^2 - \tfrac{3}{2}x^3$$

$$= -\frac{23}{2}x^3 + 7x^2 - 5x + 4 + \ldots$$

> Note that the term in x^4 is ignored as we only want the terms up to and including the term in x^3.

Expansion is valid for $|2x| < 1$ so $|x| < \tfrac{1}{2}$.

6   (a)   Explain why the sum to infinity of a geometric series with common ratio r only converges when $|r| < 1$.   [1]

(b)   A geometric progression V has first term 2 and common ratio r. Another progression W is formed by squaring each term in V. Show that W is also a geometric progression. Given that the sum to infinity of W is 3 times the sum to infinity of V, find the value of r.   [6]

(c) At the beginning of each year, a man invests £5000 in a savings account earning compound interest at the rate of 3% per annum. The interest is added at the end of each year. Find the total amount of his savings at the end of the 20th year correct to the nearest pound. [3]

(WJEC June 2018 Unit 3 q9)

Thinking about the question

In part (a) we need to understand that converging means reaching a steady value. If we look at the formula $S_n = \dfrac{a(1 - r^n)}{1 - r}$ we can see if $|r|$ is greater than 1, the value of r^n will increase as n becomes larger. If $|r|$ is less than 1, r^n will get smaller and smaller causing S_n to approach a steady value as n gets larger.

Part (b) concerns the formation of two series. If we are proving terms are in a geometric progression, then we need to show there is a common ratio between successive terms.

Starting the solution

For part (a) we need to consider that as n approaches infinity, if $|r| < 1$ the value of r^n will approach zero which means the sum will approach a steady value (i.e. it will converge).

For part (b) we need to create the terms and show there is a common ratio. We then use the formula for the sum to infinity which is included in the formula booklet.

For part (c) we are modelling the growth of an amount of money using a geometric progression. It is important to note that the £5000 is invested at the start of each year.

The solution

(a) $S_n = \dfrac{a(1 - r^n)}{1 - r}$

For the sum to infinity, the limit is $n \to \infty$

This will only converge if r^n converges.

The sum to infinity of a geometric series only converges if $|r| < 1$.

(b)   For series V terms are $2, 2r, 2r^2, ... , 2r^{n-1}$

For series W terms are $4, 4r^2, 4r^4 ..., (2r^{n-1})^2 = 4r^{2n-2}$

So, for W the nth term $= 4r^{2n-2}$

For series W, ratio $\dfrac{(\text{2nd term})}{(\text{1st term})} = \dfrac{4r^2}{4} = r^2$ and ratio $\dfrac{(\text{3rd term})}{(\text{2nd term})}$
$= \dfrac{4r^4}{4r^2} = r^2$

Hence there is a common ratio, so W is a geometric progression.

For V, sum to infinity $= \dfrac{a}{1-r} = \dfrac{2}{1-r}$

For W, sum to infinity $= \dfrac{a}{1-r} = \dfrac{4}{1-r^2}$

Now $\dfrac{4}{1-r^2} = 3 \times \dfrac{2}{1-r}$

$4(1-r) = 6(1-r^2)$

$4 - 4r = 6 - 6r^2$

$6r^2 - 4r - 2 = 0$

$3r^2 - 2r - 1 = 0$

$(3r + 1)(r - 1) = 0$

$r = -\dfrac{1}{3}$ (the other solution is ignored as $|r|<1$).

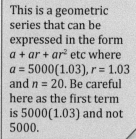

This is a geometric series that can be expressed in the form $a + ar + ar^2$ etc where $a = 5000(1.03)$, $r = 1.03$ and $n = 20$. Be careful here as the first term is $5000(1.03)$ and not 5000.

(c)   Total savings $T = 5000[(1.03) + (1.03)^2 + (1.03)^3 + ... + (1.03)^{20}]$

$S_n = \dfrac{a(r^n - 1)}{r - 1}$

$S_{20} = \dfrac{5000(1.03)(1.03^{20} - 1)}{1.03 - 1}$

$T = £138382$

We use this version of the formula as $r > 1$.

Exam practice

① Find seven numbers which are in arithmetic progression such that the last term is 71 and the sum of all of the numbers is 329.

[5]

(WJEC June 2018 Unit 3 q8)

② Find the value of $\displaystyle\sum_{n=1}^{100} (3n - 2)$.

[4]

③ If $\ln a + \ln c = 2 \ln b$, prove that the terms a, b, c form a geometric sequence.

[5]

④ Expand $(a + b)^4$. Hence expand $\left(2x + \dfrac{1}{2x}\right)^4$, simplifying each term of the expansion.

[4]

5 (a) Expand $\left(x + \dfrac{2}{x}\right)^4$, simplifying each term of the expansion. [4]

(b) The coefficients of x^2 in the expansion of $(1 + x)^n$ is 55. Given that n is a positive integer, find the value of n. [3]

(WJEC C1 May 2009 Q7)

6 In the binomial expansion of $(a + 2x)^5$, the coefficient of the term in x^2 is four times the coefficient of the term in x. Find the value of the constant a. [3]

7 (a) Expand $(1 + 4x)^{\frac{1}{2}}$ in ascending powers of x up to and including the term in x^2. State the range of values of x for which your expansion is valid. [3]

(b) Use your answer to part (a) to expand $(1 + 4y + 8y^2)^{-\frac{1}{2}}$ in ascending powers of y up to and including the term in y^2. [3]

(WJEC June 2017 C4 q5)

8 (a) Write down the expansion of $(1 + x)^6$ in ascending powers of x up to and including the term in x^3. [2]

(b) By substituting an appropriate value for x in your expansion in (a), find an approximate value for 0.99^6. Show all your working and give your answer correct to four decimal places. [3]

(WJEC C1 May 2010 Q4)

9 Write down the first three terms in the binomial expansion of $(1 - 4x)^{-\frac{1}{2}}$ in ascending powers of x. State the range of values of x for which the expansion is valid. By writing $x = \dfrac{1}{13}$ in your expansion, find an approximate value for $\sqrt{13}$ in the form $\dfrac{a}{b}$, where a, b are integers. [5]

(WJEC June 2018 Unit 3 q6)

10 (a) The eighth and ninth terms of a geometric series are 576 and 2304 respectively. Find the fifth term of the geometric series. [3]

(b) Another geometric series has first term a and common ratio r. The third term of this geometric series is 24. The sum of the second, third and fourth terms of the series is −56.

(i) Show that r satisfies the equation $3r^2 + 10r + 3 = 0$.

(ii) Given that $|r| < 1$, find the value of r and the sum to infinity of the series. [8]

(WJEC June 2015 C2 q5)

11 Find the value of, in terms of ln 2

(a) $\ln 2 + \ln (2)^2 + \ln (2)^3 + ... + \ln (2)^{50}$, [2]

(b) $\displaystyle\sum_{n=1}^{\infty} (\ln 2)^n$. [3]

> Be careful here as $\ln (2)^2$ is not the same as $(\ln 2)^2$ as you could easily be lulled into thinking this was an arithmetic progression.

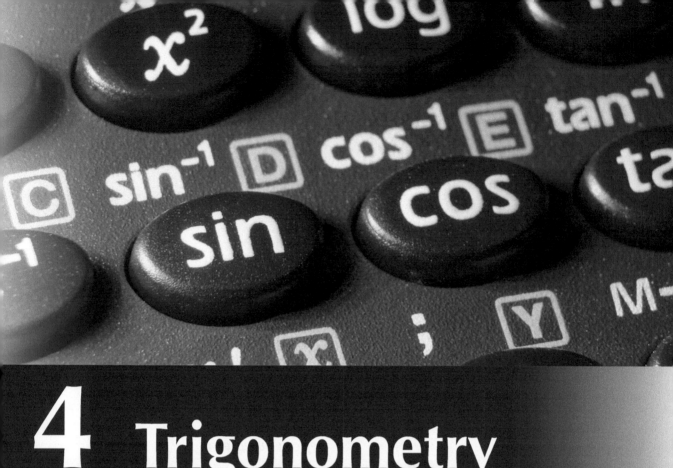

4 Trigonometry

Prior knowledge

You will need to make sure you fully understand the following from your AS studies:

- Sine, cosine and tangent and their exact values.

- Obtaining angles given a trigonometric ratio.

- Sine, cosine and tangent, their graphs, symmetries and periodicity.

- Using $\tan \theta = \frac{\sin \theta}{\cos \theta}$ and $\sin^2 \theta + \cos^2 \theta = 1$.

- Solving trigonometric equations.

Quick revision

Radian measure, arc length, area of sector and area of segment

π radians = 180°　　　　2π radians = 360°

$\dfrac{\pi}{2}$ radians = 90°　　　　$\dfrac{\pi}{4}$ radians = 45°

$\dfrac{\pi}{3}$ radians = 60°　　　　$\dfrac{\pi}{6}$ radians = 30°

The length of an arc making an angle of θ radians at the centre $l = r\theta$

Area of sector making an angle of θ radians at the centre $= \dfrac{1}{2}r^2\theta$

Area of segment $= \dfrac{1}{2}r^2(\theta - \sin\theta)$

sec, cosec and cot

$$\sec\theta = \dfrac{1}{\cos\theta}$$

$$\operatorname{cosec}\theta = \dfrac{1}{\sin\theta}$$

$$\cot\theta = \dfrac{1}{\tan\theta}$$

Trignometric identities

$$\sec^2\theta = 1 + \tan^2\theta$$

$$\operatorname{cosec}^2\theta = 1 + \cot^2\theta$$

$$\sin(A \pm B) = \sin A \cos B \pm \cos A \sin B$$

$$\cos(A \pm B) = \cos A \cos B \mp \sin A \sin B$$

$$\tan(A \pm B) = \dfrac{\tan A \pm \tan B}{1 \mp \tan A \tan B}$$

Double angle formulae

$$\sin 2A = 2\sin A \cos A$$

$$\cos 2A = \cos^2 A - \sin^2 A$$

$$= 1 - 2\sin^2 A$$

$$= 2\cos^2 A - 1$$

$$\tan 2A = \dfrac{2\tan A}{1 - \tan^2 A}$$

Important rearrangements of the double angle formulae

$$\sin^2 A = \frac{1}{2}\left(1 - \cos 2A\right)$$

$$\cos^2 A = \frac{1}{2}\left(1 + \cos 2A\right)$$

Looking at exam questions

1 Find the smallest positive root of the equation
$\sin x + \cos x + \tan x = 1.5$, where x is a small angle measured
in radians.

Give your answer correct to 3 decimal places.   [4]

Thinking about the question

As x is a small angle measured in radians, this question concerns the
small angle approximations for sine, cosine and tangent. As none of the
formulae for small angle approximations are included in the formula
booklet, they will need to be recalled from memory.

Starting the solution

The identities $\sin x \approx x$, $\cos x \approx 1 - \frac{1}{2}x^2$ and $\tan x \approx x$ need to be recalled
and then substituted into the equation. As there is a term in x^2, this
usually results in a quadratic equation, which will need to be solved
using the formula as the question requires the angle to be given to
3 d.p. Notice also that only the positive root is required.

The solution

For small angles measured in radians, we have the following
identities:

$\sin x \approx x$ and $\cos x \approx 1 - \frac{1}{2}x^2$

$\tan x \approx x$

$\sin x + \cos x + \tan x = 1.5$

$x + 1 - \frac{1}{2}x^2 + x \approx 1.5$

$\frac{1}{2}x^2 - 2x + 0.5 \approx 0$

$x^2 - 4x + 1 = 0$

$x = \dfrac{4 \pm \sqrt{16 - 4}}{2}$

$= \dfrac{4 \pm \sqrt{12}}{2}$

$= 3.732$ or 0.268

We require the smallest positive root. Hence $x = 0.268$ rads (3 d.p.)

> **Note** that these are not included in the formula booklet and so need to be remembered.

> **Watch out**
>
> The question asks for the solution to be given to 3 decimal places so don't waste time trying to factorise the quadratic equation. Instead use the formula.

2 The diagram below shows a circle centre O, radius 4 cm. Points A and B lie on the circumference such that arc AB is 5 cm.

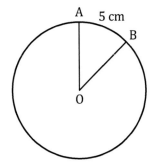

(a) Calculate the angle subtended at O by the arc AB. [2]

(b) Determine the area of the sector OAB. [2]

(WJEC June 2018 Unit 3 q2)

Thinking about the question

When we are dealing with circles we usually measure the angle in radians, although the question does not specify what units the angle is measured in. We need to recall the formulae for the length of an arc and the area of a sector as they are not included in the formula booklet.

Starting the solution

For part (a) we recall and use the formula $l = r\theta$ to work out the angle θ in radians.

For part (b) we recall and use the formula $A = \dfrac{1}{2}r^2\theta$ to work out A.

The solution

(a) Length of arc $= r\theta$

$$5 = 4\theta$$

$$\theta = \frac{5}{4} \text{ radians}$$

(b) Area $= \dfrac{1}{2}r^2\theta$

$$= \frac{1}{2} \times 4^2 \times \frac{5}{4}$$

$$= 10 \text{ cm}^2$$

3   (a)   Find all the values of θ in the range $0° \le \theta \le 360°$ satisfying

$$2 \cos 2\theta = 9 \cos \theta + 7 \qquad [5]$$

   (b)   (i)   Express $5 \sin x - 12 \cos x$ in the form $R \sin (x - \alpha)$, where R and α are constants with $R > 0$ and $0° < \alpha < 90°$.

   (ii)   Use your results to part (i) to find the least value of

$$\frac{1}{5 \sin x - 12 \cos x + 20}.$$

Write down a value for x for which this least value occurs. $\qquad [6]$

(WJEC C4 June 2010 Q3)

Thinking about the question

For part (a) we need to expand $\cos 2\theta$ and then form an equation that can be solved.

For part (b)(i) we need to expand $R \sin (x - \alpha)$ using the formula for the expansion of $\sin (A - B)$ which is included in the formula booklet and then compare the answer with the given expression.

For part (b)(ii) we use the result from (b)(i) as this is in the denominator of the fraction. For the smallest value of the fraction the denominator needs to have its largest value. As the largest value of a sin function is 1, we need to find the value of x for this to occur.

Starting the solution

For part (a) we obtain the formula $\cos (A + B) = \cos A \cos B - \sin A \sin B$ and use $B = A$ to give $\cos (A + A) = \cos A \cos A - \sin A \sin A = \cos^2 A - \sin^2 A$. Since $\sin^2 A + \cos^2 A = 1$ so $\sin^2 A = 1 - \cos^2 A$. We can now write $\cos 2A = 2 \cos^2 A - 1$.

As the equation is now a quadratic equation it can be factorised and then solved.

The solution

(a) $$2 \cos 2\theta = 9 \cos \theta + 7$$

$$2(2 \cos^2 \theta - 1) = 9 \cos \theta + 7$$

$$4 \cos^2 \theta - 2 = 9 \cos \theta + 7$$

$$4 \cos^2 \theta - 9 \cos \theta - 9 = 0$$

$$(4 \cos \theta + 3)(\cos \theta - 3) = 0$$

Hence $\cos \theta = -\dfrac{3}{4}$ or $\cos \theta = 3$ (This value is ignored as max value of a cos function is 1).

When $\cos \theta = -\dfrac{3}{4}$, $\theta = 138.59°, 221.41°$

(b) (i)　$5 \sin x - 12 \cos x \equiv R \sin(x - \alpha)$

$5 \sin x - 12 \cos x \equiv R \sin x \cos \alpha - R \cos x \sin \alpha$

$R \cos \alpha = 5$ and $R \sin \alpha = 12$

So $\tan \alpha = \dfrac{12}{5}$ giving $\alpha = 67.38°$

$R = \sqrt{5^2 + 12^2} = \sqrt{169} = 13$

Hence $5 \sin x - 12 \cos x = 13 \sin(x - 67.38°)$

> Use $\sin(A \pm B) = \sin A \cos B \pm \cos A \sin B$ obtained from the formula booklet.

(b) (ii)　$\dfrac{1}{5 \sin x - 12 \cos x + 20} = \dfrac{1}{13 \sin(x - 67.38°) + 20}$

For this expression to have its least value the denominator must have its largest value.

This would occur when $\sin(x - 67.38°) = 1$ in which case:

$\dfrac{1}{13 \sin (x - 67.38°) + 20} = \dfrac{1}{13 + 20} = \dfrac{1}{33}$

Now $\sin(x - 67.38°) = 1$ so $x - 67.38° = 90°$ giving $x = 157.38°$

---

4 (a) Given that α and β are two angles such that $\tan \alpha = 2 \cot \beta$, show that:

$$\tan (\alpha + \beta) = -(\tan \alpha + \tan \beta). \qquad [2]$$

(b) Find all values of θ in the range $0° \leq \theta \leq 360°$ satisfying the equation:

$$4 \tan \theta = 3 \sec^2 \theta - 7. \qquad [6]$$

(WJEC Unit 3 June 2019 q9)

Thinking about the question

This question involves solving trigonometric equations, so we need to refer to the formula sheet for some trig formulae as well as recall simple ones ourselves.

Starting the solution

For part (a) we need to obtain the expansion of $\tan(A + B)$ from the formula booklet and then recall the fact that $\cot \beta = \dfrac{1}{\tan \beta}$. By combining these we will arrive at the correct result.

For part (b) we need to create a quadratic equation in $\tan \theta$ using the fact that $\sec^2 \theta = 1 + \tan^2 \theta$ which we then solve for values of $\tan \theta$. We then find the angles θ in the required range.

The solution

Notice the numerator is $\tan \alpha + \tan \beta$ and this is in the result you want to prove. So it is best to leave the numerator alone and only substitute $\tan \alpha = 2 \cot \beta$ into the denominator.

(a)
$$\tan(\alpha + \beta) = \frac{\tan \alpha + \tan \beta}{1 - \tan \alpha \tan \beta}$$

$$= \frac{\tan \alpha + \tan \beta}{1 - 2 \cot \beta \tan \beta}$$

Now
$$\cot \beta = \frac{1}{\tan \beta}$$

Hence, $\tan(\alpha + \beta) = \dfrac{\tan \alpha + \tan \beta}{1 - \dfrac{2}{\tan \beta} \tan \beta}$

$$= \frac{\tan \alpha + \tan \beta}{-1}$$

$$= -(\tan \alpha + \tan \beta)$$

(b)
$$4 \tan \theta = 3 \sec^2 \theta - 7$$

Now
$$\sec^2 \theta = 1 + \tan^2 \theta$$

$$4 \tan \theta = 3(1 + \tan^2 \theta) - 7$$

$$4 \tan \theta = 3 + 3 \tan^2 \theta - 7$$

$$3 \tan^2 \theta - 4 \tan \theta - 4 = 0$$

$$(3 \tan \theta + 2)(\tan \theta - 2) = 0$$

$$\tan \theta = -\frac{2}{3} \text{ or } \tan \theta = 2$$

When $\tan \theta = -\dfrac{2}{3}$, $\theta = 146.31°$ or $326.31°$

When $\tan \theta = 2$, $\theta = 63.43°$ or $243.43°$

---

5   Solve the equation $2 \tan^2 \theta + 2 \tan \theta - \sec^2 \theta = 2$,

for values of θ between $0°$ and $360°$.   [5]

(WJEC Unit 3 June 2018 q4)

Thinking about the question

Look carefully at the equation. It looks like the start of a quadratic equation in $\tan \theta$ but there is a term in $\sec^2 \theta$. We need to change this $\sec^2 \theta$.

Starting the solution

We start by using $\sec^2 \theta = 1 + \tan^2 \theta$ and substituting this into the equation, which is then rearranged to form a quadratic which can be factorised and solved.

The solution

$$2 \tan^2 \theta + 2 \tan \theta - \sec^2 \theta = 2$$

Now $$\sec^2 \theta = 1 + \tan^2 \theta$$

$$2 \tan^2 \theta + 2 \tan \theta - (1 + \tan^2 \theta) = 2$$

$$2 \tan^2 \theta + 2 \tan \theta - 1 - \tan^2 \theta = 2$$

$$\tan^2 \theta + 2 \tan \theta - 3 = 0$$

$$(\tan \theta + 3)(\tan \theta - 1) = 0$$

$$\tan \theta = 1 \text{ or } \tan \theta = -3$$

When $\tan \theta = -3$, $\theta = 108.43°, 288.43°$

When $\tan \theta = 1$, $\theta = 45°, 225°$

6

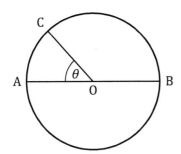

The diagram shows three points A, B, C on a circle with centre O and radius 4 cm, such that AB is a diameter of the circle and $\hat{AOC} = \theta$ radians. Given that the area of the sector BOC is 5 cm² more than the area of the sector AOC,

(a) Show that $\theta = \dfrac{8\pi - 5}{16}$. [3]

(b) Calculate the difference between the arc length BC and the arc length AC. [3]

(WJEC June 2005 C2 q9)

Thinking about the question

This question concerns areas of sectors when the angle is in radians. We need to recall the formula $A = \dfrac{1}{2} r^2 \theta$ from memory. As the angle of the sector AOC is θ, the angle BOC is $\pi - \theta$.

For part (b) we need to recall the formula for the length of an arc, $l = r\theta$.

Starting the solution

We need to create two equations for the areas of each of the sectors and then form a single equation which can be rearranged to the given equation.

For part (b) we use the arc length formula to find an equation for the difference in the arc lengths. We can then substitute the equation for θ found in part (a) into this equation.

The solution

(a)   Area of sector AOC $= \dfrac{1}{2}r^2\theta = \dfrac{1}{2} \times 4^2 \times \theta = 8\theta$

Area of sector BOC $= \dfrac{1}{2}r^2\theta = \dfrac{1}{2} \times 4^2 \times (\pi - \theta) = 8(\pi - \theta)$

Now, $8(\pi - \theta) = 8\theta + 5$

$$8\pi - 8\theta = 8\theta + 5$$

$$8\pi - 5 = 16\theta$$

$$\theta = \frac{8\pi - 5}{16}$$

(b)   Length of arc AC $= r\theta = 4\theta$

Length of arc BC $= 4(\pi - \theta)$

Difference in arc length between
BC and AC $= 4(\pi - \theta) - 4\theta = 4\pi - 8\theta$

$$= 4\pi - 8\left(\frac{8\pi - 5}{16}\right) = 4\pi - \left(\frac{8\pi - 5}{2}\right) = \frac{5}{2} \text{ cm}$$

> Notice the question asks you to calculate the arc length so we are required to give it as a number.

Exam practice

1.   Use small angle approximations to find the negative root of the equation:
$$\sin x + \cos x = 0.5 \tag{[5]}$$
(WJEC Unit 3 June 2018 q7)

2.   Find all the values of θ in the range $0° \leq \theta \leq 360°$ satisfying
$3 \sin 2\theta = 2 \sin \theta$.   [5]
(WJEC C4 June 2009 Q2)

3.   Find the values of A between $0°$ and $360°$ satisfying:
$$\tan 2A = 3 \tan A \tag{[5]}$$

4.   (a) Express $3 \cos \theta - 4 \sin \theta$ in the form $R \cos(\theta + \alpha)$ where $R > 0$ and $0° < \alpha < 90°$ [3]
   (b) Solve $3 \cos \theta - 4 \sin \theta = 2.5$ for values of θ between $0°$ and $360°$ [3]

5.   (a) Express $2 \cos^2 \theta + 6 \sin \theta \cos \theta$ in the form $A + B \cos 2\theta + C \sin 2\theta$ [3]
   (b) Solve the equation $2 \cos^2 \theta + 6 \sin \theta \cos \theta = 2$ for values of θ between $0°$ and $180°$.   [5]

6

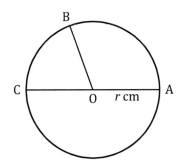

The diagram shows a sketch of a circle with centre O and radius r cm. Three points A, B and C lie on the circle. The line AC is a diameter of the circle and $A\hat{O}B = 2{\cdot}15$ radians. Given that the area of sector BOC is 26 cm² less than the area of sector AOB, find the value of r. Give your answer correct to one decimal place. [5]

(WJEC Core 2 May 2016 q9)

7 (a) Express $8 \sin \theta - 15 \cos \theta$ in the form $R \sin(\theta - \alpha)$, where R and α are constants with $R > 0$ and $0° < \alpha < 90°$. [3]

   (b) Find all values of θ in the range $0° < \theta < 360°$ satisfying:
$$8 \sin \theta - 15 \cos \theta - 7 = 0.$$
[3]

   (c) Determine the greatest value and the least value of the expression:
$$\frac{1}{8 \sin \theta - 15 \cos \theta + 23}.$$
[2]

(WJEC Unit 3 June 2018 q13)

8 Find the values of θ in the range $0° < \theta < 360°$ that satisfy the equation:
$$4 \sec^2 \theta = 7 - 11 \tan \theta$$
giving your answer to 1 decimal place. [5]

9

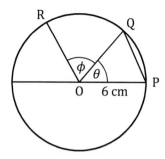

The diagram shows three points P, Q and R on a circle with centre O. The circle has radius 6 cm, POQ = θ radians and QOR = ϕ radians.

(a) The area of the *triangle* POQ is 9·1 cm². Find the value of θ, giving your answer correct to two decimal places.   [2]

(b) Find the area of the sector POQ. Give your answer correct to two decimal places.   [2]

(c) The **perimeter** of the **sector** QOR is equal to half the circumference of the circle. Find the value of ϕ, giving your answer correct to two decimal places.   [2]

(WJEC Jan 2010 C2 q9)

5 Differentiation

Prior knowledge

You will need to make sure you fully understand the following from your AS studies:

- Simple differentiation.
- Finding stationary points and their nature.
- Simple curve sketching.

Quick revision

Formulae

Differentiation of e^{kx}, a^{kx}, $\sin kx$, $\cos kx$ and $\tan kx$

> All these derivatives need to be remembered.

$$\frac{d(e^{kx})}{dx} = ke^{kx}$$

$$\frac{d(a^{kx})}{dx} = ka^{kx}\ln a$$

$$\frac{d(\sin kx)}{dx} = k\cos kx$$

$$\frac{d(\cos kx)}{dx} = -k\sin kx$$

The following derivative need not be remembered as it is included in the formula booklet.

$$\frac{d(\tan kx)}{dx} = k\sec^2 kx$$

The derivative of $\ln x$

> This derivative needs to be remembered.

$$\frac{d(\ln x)}{dx} = \frac{1}{x}$$

To differentiate the natural logarithm of a function you differentiate the function and then divide by the function.

This can be expressed mathematically as:

$$\frac{d(\ln (f(x)))}{dx} = \frac{f'(x)}{f(x)}$$

The Chain rule

If y is a function of u and u is a function of x, then the Chain rule states:

$$\frac{dy}{dx} = \frac{dy}{du} \times \frac{du}{dx}$$

The Product rule

If $y = f(x)g(x)$, $\dfrac{dy}{dx} = f(x)g'(x) + g(x)f'(x)$

The Quotient rule

If $y = \dfrac{f(x)}{g(x)}$, $\dfrac{dy}{dx} = \dfrac{f'(x)g(x) - f(x)g'(x)}{(g(x))^2}$

Differentiation of inverse functions $\sin^{-1} x$, $\cos^{-1} x$, $\tan^{-1} x$

$$\frac{d(\sin^{-1} x)}{dx} = \frac{1}{\sqrt{1 - x^2}}$$

$$\frac{d(\cos^{-1} x)}{dx} = -\frac{1}{\sqrt{1 - x^2}}$$

$$\frac{d(\tan^{-1} x)}{dx} = -\frac{1}{1 + x^2}$$

Differentiation of simple functions defined implicitly

Finding $\frac{dy}{dx}$ in terms of both x and y is called implicit differentiation.

Here are the rules for differentiating implicitly:

- Terms involving x or constant terms are differentiated as normal.
- For terms just involving y, (e.g. $3y$, $5y^3$, etc.) differentiate with respect to y and then multiply the result by $\frac{dy}{dx}$.
- For terms involving both x and y (e.g. xy, $5x^2y^3$, etc.) the Product rule is used because there are two terms multiplied together. Note the need to include $\frac{dy}{dx}$ when the term involving y is differentiated.

Differentiation of functions defined parametrically

The equation of a curve can be expressed in parametric form by using:

$$x = f(t), y = g(t) \quad \text{where } t \text{ is the parameter being used.}$$

The formulae for differentiating parametric forms are:

$$\frac{dy}{dx} = \frac{\frac{dy}{dt}}{\frac{dx}{dt}} = \frac{dy}{dt} \times \frac{dt}{dx}$$

and

$$\frac{d^2y}{dx^2} = \frac{\frac{d}{dt}\left(\frac{dy}{dx}\right)}{\frac{dx}{dt}} = \frac{d}{dt}\left(\frac{dy}{dx}\right) \times \frac{dt}{dx}$$

Looking at exam questions

1 Curve C is given by the equation $y = xe^x$.

   (a) Find the coordinates of the turning point on C and give its nature. [4]

   (b) Curve C also has a point of inflection. Find the coordinates of the point of inflection and draw a sketch of curve C. [4]

Thinking about the question

This question is about turning points and points of inflection, so we need to find the first and then the second derivative of the equation of C.

Starting the solution

For part (a) we need to differentiate the equation of C and then equate it to zero and solve to find the turning point. We notice that there is only a single turning point. On finding the x-coordinate we need to substitute it into the equation of C to find the y-coordinate. To find the nature of this point we find the second derivative and substitute our value in for x to see if it results in either a positive (for a min) or a negative (for a max) value.

For part (b) we use the second derivative and equate it to zero and then solve to find the value of x. We then investigate the value of the second derivative at points on either side and near to the value of x. A change in sign indicates a point of inflection. We then need to substitute the value of x into the equation of C to find the y-coordinate of the point.

We then need to draw a sketch of the curve. We already know the minimum point and the point of inflection, so we just need to find out the y-value when $x = 0$ and also any asymptotes.

The solution

> This is a product, so we use the Product rule for the differentiation. You must remember the Product rule.

(a) $y = xe^x$

$$\frac{dy}{dx} = xe^x + e^x(1) = e^x(x + 1)$$

At the turning point, $\frac{dy}{dx} = 0$, so $e^x(x + 1) = 0$

Now, there is no value of x for which $e^x = 0$

Hence $(x + 1) = 0$, so $x = -1$

When $x = -1$, $y = xe^x = (-1)e^{-1} = -\dfrac{1}{e}$

There is a stationary point at $\left(-1, -\dfrac{1}{e}\right)$

> The second derivative is found so that the nature (i.e. maximum or minimum) of the stationary point can be determined.

$$\frac{d^2y}{dx^2} = e^x(1) + (x + 1)e^x$$

$$= e^x + xe^x + e^x$$

$$= 2e^x + xe^x$$

$$= e^x(2 + x)$$

When $x = -1$, $\dfrac{d^2y}{dx^2} = e^{-1}(2 - 1) = \dfrac{1}{e}$ (which is positive)

Hence $\left(-1, -\dfrac{1}{e}\right)$ is a minimum point.

(b) For the point of inflection, $\dfrac{d^2y}{dx^2} = 0$ so $e^x(2+x) = 0$

Now there is no value of x for which $e^x = 0$.

Hence $2 + x = 0$ so $x = -2$

When x is slightly less than -2, $\dfrac{d^2y}{dx^2}$ is positive and when x is slightly more than -2, $\dfrac{d^2y}{dx^2}$ is negative so $\dfrac{d^2y}{dx^2}$ changes sign.

When $x = -2$, $y = xe^x = -2e^{-2}$

Hence there is a point of inflection at $\left(-2, -\dfrac{2}{e^2}\right)$

$y = xe^x$ so when $x = 0$, $y = 0$ so the curve passes through the origin.

When $x \to \infty$, $y \to \infty$

When $x \to -\infty$, $y \to 0$

Adding all this information and the coordinates of the minimum point and the point of inflection, we obtain the graph:

> Remember $e^0 = 1$.

> The x-coordinate is substituted into the equation of the curve to find the y-coordinate.

> Consider what happens to y when x becomes a very large positive value. Then consider what happens when x becomes a very large negative value.

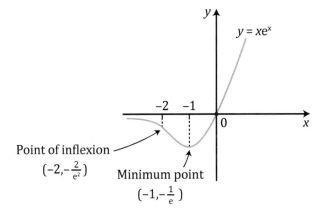

2 Given that $x = \ln t$ and $y = 2t^2 - t$.

(a) Find an expression for $\dfrac{dy}{dx}$ in terms of t, [3]

(b) Find the value of $\dfrac{d^2y}{dx^2}$ when $t = \dfrac{1}{2}$. [3]

Thinking about the question

This is a question about parametric equations with x and y both expressed in terms of a parameter, t.

Starting the solution

For part (a) we need to find $\dfrac{dx}{dt}$ and $\dfrac{dy}{dt}$ and then combine them to give $\dfrac{dy}{dx}$. We remember that the differential of $\ln x$ is $\dfrac{1}{x}$.

For part (b) we need to find the second derivative and then substitute in $t = \dfrac{1}{2}$.

5 Differentiation

The solution

(a) $x = \ln t$

$\dfrac{dx}{dt} = \dfrac{1}{t}$

$y = 2t^2 - t$

$\dfrac{dy}{dt} = 4t - 1$

$\dfrac{dy}{dx} = \dfrac{dy}{dt} \times \dfrac{dt}{dx}$

$\dfrac{dy}{dx} = (4t - 1)t$

$\dfrac{dy}{dx} = 4t^2 - t$

$\dfrac{dt}{dx} = \dfrac{1}{\frac{1}{t}} = t$

The first derivative is differentiated again to find the second derivative. Note that you need to remember that $\dfrac{d^2y}{dx^2} = \dfrac{d}{dt}\left(\dfrac{dy}{dx}\right) \times \dfrac{dt}{dx}$

(b) $\dfrac{d^2y}{dx^2} = \dfrac{d}{dt}(4t^2 - t)\dfrac{dt}{dx}$

$= (8t - 1)t$

$= 8t^2 - t$

When $t = \dfrac{1}{2}, \dfrac{d^2y}{dx^2} = 8\left(\dfrac{1}{2}\right)^2 - \dfrac{1}{2} = 2 - \dfrac{1}{2} = 1.5$

---

3 (a) Differentiate each of the following functions with respect to x.

(i) $x^5 \ln x$

(ii) $\dfrac{e^{3x}}{x^3 - 1}$

(iii) $(\tan x + 7x)^{\frac{1}{2}}$ [8]

(b) A function is defined implicitly by:

$3y + 4xy^2 - 5x^3 = 8$

Find $\dfrac{dy}{dx}$ in terms of x and y. [3]

(WJEC June 2019 unit 3 q10)

Thinking about the question

Part (a) covers the various methods of differentiation. Part (a)(i) is a product, (ii) is a quotient and (iii) is a function of a function.

For part (b) we need to use implicit differentiation.

Starting the solution

(a) (i) Is a product so we need to use the Product rule, which needs to be remembered as it is not included in the formula booklet. It can be remembered as the first term multiplied by the derivative of the second term plus the second term multiplied by the derivative of the first.

(ii) Is a quotient so we need to look up the formula in the formula booklet.

The formula for differentiating $y = \dfrac{f(x)}{g(x)}$ is $\dfrac{dy}{dx} = \dfrac{f'(x)\,g(x) - f(x)\,g'(x)}{(g(x))^2}$

(iii) Is a function of a function so we let u = the contents of the bracket.

The solution

(a) (i) Let $y = x^5 \ln x$

$$\frac{dy}{dx} = x^5\left(\frac{1}{x}\right) + \ln x\,(5x^4)$$

$$= x^4 + 5x^4 \ln x$$

$$= x^4\,(1 + 5\ln x)$$

> The Product rule is used here.

(ii) Let $y = \dfrac{e^{3x}}{x^3 - 1}$

$$\frac{dy}{dx} = \frac{(x^3 - 1)3e^{3x} - e^{3x}(3x^2)}{(x^3 - 1)^2}$$

$$= \frac{3x^3 e^{3x} - 3e^{3x} - 3x^2 e^{3x}}{(x^3 - 1)^2}$$

$$= \frac{3e^{3x}\,(x^3 - x^2 - 1)}{(x^3 - 1)^2}$$

> This is a quotient, so the Quotient rule is used, which is included in the formula booklet.

(iii) Let $y = (\tan x + 7x)^{\frac{1}{2}}$

Let $u = \tan x + 7x$ so $y = u^{\frac{1}{2}}$ and $\dfrac{dy}{du} = \dfrac{1}{2}u^{-\frac{1}{2}}$

$$\frac{du}{dx} = \sec^2 x + 7$$

$$\frac{dy}{dx} = \frac{dy}{du} \times \frac{du}{dx}$$

$$= (\sec^2 x + 7)\left(\frac{1}{2}u^{-\frac{1}{2}}\right)$$

$$= -\frac{1}{2}(\sec^2 x + 7)(\tan x + 7x)^{-\frac{1}{2}}$$

> This is a function of a function, so we let $u = \tan x + 7x$.

> We now replace u with $\tan x + 7x$

(b) $3y + 4xy^2 - 5x^3 = 8$

Differentiating implicitly with respect to x, we obtain:

$$3\frac{dy}{dx} + 4x(2y)\frac{dy}{dx} + y^2(4) - 15x^2 = 0$$

$$\frac{dy}{dx}(3 + 8xy) = 15x^2 - 4y^2$$

$$\frac{dy}{dx} = \frac{15x^2 - 4y^2}{3 + 8xy}$$

> Note that to differentiate $4xy^2$ we use the Product rule.

---

4 Find the coordinates of the point of inflection on the curve
$y = x(x - 3)(x + 4)$. [6]

Thinking about the question

This question involves finding the first and second derivatives. For a point of inflection, the second derivative is zero.

Starting the solution

We first start off by multiplying out the brackets so that the equation is in a form that is easy to differentiate. We then find the first derivative and then differentiate this to give the second derivative. Equating the second derivative to zero and solving the resulting equation will give us the x-coordinate of the point of inflection. We can then substitute the x-coordinate back into the equation of the curve to give the y-coordinate.

We then have to verify that that the point is a point of inflection by looking at the sign of the second derivative either side of the x-coordinate we have found. If there is a change in sign, the point is a point of inflection.

The solution

$$y = x(x - 3)(x + 4)$$

$$= x(x^2 + 4x - 3x - 12)$$

$$= x^3 + x^2 - 12x$$

$$\frac{dy}{dx} = 3x^2 + 2x - 12$$

$$\frac{d^2y}{dx^2} = 6x + 2$$

> At a point of inflection, the second derivative is equal to zero.

When $\frac{d^2y}{dx^2} = 0$, $6x + 2 = 0$, so $x = -\frac{1}{3}$

When $x = -\frac{1}{2}, \frac{d^2y}{dx^2} = 6\left(-\frac{1}{2}\right) + 2 = -1$

> Here we check there is a change in sign for the second derivative either side of the point with x-coordinate $-\frac{1}{3}$.

When $x = -\frac{1}{6}, \frac{d^2y}{dx^2} = 6\left(-\frac{1}{6}\right) + 2 = 1$

There is a sign change so $x = -\frac{1}{3}$ is a point of inflection.

When $x = -\frac{1}{3}, y = x^3 + x^2 - 12x = \left(-\frac{1}{3}\right)^3 + \left(-\frac{1}{3}\right)^2 - 12\left(-\frac{1}{3}\right) = 4\frac{2}{27}$

> Don't forget to work out the y-coordinate.

Hence, point of inflection is at $\left(-\frac{1}{3}, 4\frac{2}{27}\right)$

Exam practice

1. Differentiate each of the following:
   (a) $y = \ln 3x$ [2]
   (b) $y = \sqrt{1 - 3x^2}$ [2]
   (c) $y = x^3 \cos 3x$ [2]

> Here we use the laws of logarithms to separate $\ln 3x$ into $\ln x + \ln 3$.

2. Differentiate each of the following:
   (a) $y = \tan x^2$ [2]
   (b) $y = e^{2x} \sin 2x$ [2]

3. If $y = \ln (t^2 + 2)$ and $x = 2t^3 + t^2$, find $\dfrac{dy}{dx}$. [3]

4. Find $\dfrac{dy}{dx}$ for the equation:
   $$5x^3 + 2x^2 y + y^3 = 10$$ [4]

5. Differentiate $y = \sin (3x^3 + 2x^2 + x - 1)$ [3]

6. Differentiate $y = \ln (3x^2 + 4)$ [3]

> There are two functions here: the contents of the bracket and the sine function.

7. (a) A function is defined implicitly by
   $$x^2 + 2xy + 3y^2 = 12$$
   Find $\dfrac{dy}{dx}$ in terms of x and y. [3]

   (b) Another function is defined parametrically by $x = 2t^4, y = 3t^2$.

   (i) Find $\dfrac{dy}{dx}$ in terms of t.

   (ii) Find $\dfrac{d^2y}{dx^2}$ in terms of t. [4]

   (WJEC C3 June 2005 q4)

8. Differentiate each of the following, simplifying your answer wherever possible:
   (a) $3x^2 \tan 4x$ [2]
   (b) $\cos^{-1} 2x$ [2]
   (c) $\dfrac{\ln x}{x^3}$ [3]

9. The radius of an oil slick on the surface of the sea is increasing at a rate of 1.5 metres per minute. Find the rate of change of area when the radius is 100 m. Give your answer in metres^2 per minute. [5]

10. If $y = \sin^{-1}\left(\dfrac{1}{x}\right)$, show that $\dfrac{dy}{dx} = -\dfrac{1}{x\sqrt{x^2 - 1}}$. [4]

11. If $y = (x^2 + 1) \tan^{-1} x$, find $\dfrac{dy}{dx}$. [3]

12. If $y = \cos^{-1} (3x^2)$, find $\dfrac{dy}{dx}$. [3]

13. Find the equation of the tangent to the curve
    $x^3 + 4xy - 2x - y^3 + 1 = 0$ at point P(1, 2). [6]

14. Find the coordinates of the points of inflection on the curve
    $$y = x^4 - 6x^2 + 2$$ [7]

6 Coordinate geometry in the (x, y) plane

Prior knowledge

You will need to make sure you fully understand the following from your AS studies:

- Finding the equation of a tangent and normal to a curve.

Quick revision

Cartesian equations connect x and y in some way. For example, $y = 4x^3$ is a Cartesian equation.

Parametric equations express x and y in terms of a parameter such as t, for example:

$$x = 4 + 2t \qquad y = 1 + 2t$$

To obtain the Cartesian equation from the parametric equation it is necessary to eliminate the parameter.

$$\text{Note that} \quad \frac{dy}{dx} = \frac{dy}{dt} \times \frac{dt}{dx}$$

Using the Chain rule to find the second derivative

The Chain rule can be used to find the second derivative in terms of a parameter such as t in the following way:

$$\frac{d^2y}{dx^2} = \frac{d}{dx}\left(\frac{dy}{dx}\right) = \frac{d}{dt}\left(\frac{dy}{dx}\right)\frac{dt}{dx}$$

Implicit differentiation

Here are the basic rules:

$$\frac{d(3x^2)}{dx} = 6x$$

Terms involving x or constant terms are differentiated as normal.

$$\frac{d(6y^3)}{dx} = 18y^2 \times \frac{dy}{dx}$$

Differentiate with respect to y and then multiply the result by $\frac{dy}{dx}$.

$$\frac{d(y)}{dx} = 1 \times \frac{dy}{dx}$$

When you are differentiating a term just involving y, you differentiate with respect to y and then multiply the result by $\frac{dy}{dx}$. This is an application of the Chain rule.

$$\frac{d(x^2y^3)}{dx} = (x^2)\left(3y^2 \times \frac{dy}{dx}\right) + (y^3)(2x)$$

$$= 3x^2y^2\frac{dy}{dx} + 2xy^3$$

Because there are two terms here, the Product rule is used. Notice the need to include $\frac{dy}{dx}$ when the term involving y is differentiated.

Looking at exam questions

1   Line *L* has the parametric equations $x = 6 - 3t$ and $y = 6t - 3$.

Find the Cartesian equation of *L*.   [4]

Thinking about the question

Here we eliminate the parameter, *t*, so we have an equation in *x* and *y*.

Starting the solution

Start by rearranging one of the equations (we have used the one for *x* here) for *t*.

Then substitute *t* into the other equation so that you have an equation just connecting *x* and *y*.

The solution

$$x = 6 - 3t$$

$$3t = 6 - x$$

Substituting this into the equation $y = 6t - 3$, we obtain:

$$y = 2(6 - x) - 3$$

$$y = 12 - 2x - 3$$

$$y = -2x + 9$$

---

2   Find the equation of the normal to the curve:

$$2x^2 + 3x^2 y - 8 = 0$$

at the point (1, 2) giving your equation in the form $ax + by + c = 0$.
                                                                    [5]

Thinking about the question

As the equation of the curve has terms in both *x* and *y*, this will have to be differentiated implicitly to find the gradient of the tangent.

Starting the solution

First, we differentiate the equation of the curve implicitly. We notice that $3x^2 y$ will have to be differentiated using the Product rule which we need to recall from memory.

We can substitute the coordinates of the point to give a value for the gradient. We use $m_1 m_2 = -1$ to work out the gradient of the normal and then substitute this with the coordinates (1, 2) into $y - y_1 = m(x - x_1)$ to give the equation of the normal. We need to rearrange this equation so it is in the form asked for in the question.

The solution

$2x^2 + 3x^2 y - 8 = 0$

Differentiating implicitly with respect to x, we obtain:

$4x + 3x^2 (1)\dfrac{dy}{dx} + y(6x) = 0$

$4x + 3x^2\dfrac{dy}{dx} + 6xy = 0$

$\dfrac{dy}{dx} = \dfrac{-6y - 4}{3x}$

> We divide the numerator and denominator by x.

At the point $(1, 2)$, $\dfrac{dy}{dx} = \dfrac{-16}{3}$

Gradient of normal at $(1, 2) = \dfrac{3}{16}$

> To find the gradient of the normal we invert the fraction and change the sign.

Equation of normal is:

$$y - 2 = \dfrac{3}{16}(x - 1)$$

$$16y - 32 = 3x - 3$$

$$3x - 16y + 29 = 0$$

---

3 The equation of curve C has the parametric equations $x = 3\sin 4t$ and $y = 4\cos 4t$.

Show that the equation of the tangent to C at the point P with parameter p is
$$3y + 4x\tan 4p - 12\cos 4p - 12\tan 4p\sin 4p = 0. \qquad [6]$$

Thinking about the question

For this question we need to find the gradient at P and then use the coordinates of P to find the equation of the tangent.

Starting the solution

We find $\dfrac{dx}{dt}$ and $\dfrac{dy}{dt}$ and then use the Chain rule to find $\dfrac{dy}{dx}$. At P the parameter is p, so we substitute this into the expression for the gradient. We then use the equation for a straight line and substitute in the gradient and the x and y coordinates remembering to change the parameters from t to p.

Coordinate geometry in the (x, y) plane

Note that the derivatives of sin and cos are not included in the formula booklet.

Remember that
$\frac{\sin 4t}{\cos 4t} = \tan 4t$

The formula for a straight line is used here
(i.e. $y - y_1 = m(x - x_1)$).

The solution

$$x = 3 \sin 4t$$

$$\frac{dx}{dt} = 12 \cos 4t$$

$$y = 4 \cos 4t$$

$$\frac{dy}{dt} = -16 \sin 4t$$

$$\frac{dy}{dx} = \frac{dy}{dt} \times \frac{dt}{dx}$$

$$= \frac{-16 \sin 4t}{12 \cos 4t}$$

$$= -\frac{4}{3} \tan 4t$$

Equation at tangent at P is given by:

$$y - 4 \cos 4p = -\frac{4}{3} \tan 4p(x - 3 \sin 4p)$$

$$3y - 12 \cos 4p = -4x \tan 4p + 12 \tan 4p \sin 4p$$

$$3y + 4x \tan 4p - 12 \cos 4p - 12 \tan 4p \sin 4p = 0$$

---

4 The equation of a curve C is given by the parametric equations $x = \cos 2\theta, y = \cos \theta$.

(a) Find the Cartesian equation of C. [2]

(b) Show that the line $x - y + 1 = 0$ meets C at the point P, where $\theta = \frac{\pi}{3}$, and at the point Q, where $\theta = \frac{\pi}{2}$. Write down the coordinates of P and Q. [5]

(c) Determine the equations of the tangents to C at P and Q. Write down the coordinates of the point of intersection of the two tangents. [7]

(WJEC June 2018 unit 3 q10)

Thinking about the question

For part (a) we need to find an equation connecting x and y. For part(b) we need to solve the equation of the curve with the equation of the line. For part (c) we need to find the gradient of the curve at points P and Q and then use the coordinates to find the equations of the tangents.

Starting the solution

For part (a) we need to expand $x = \cos 2\theta$ and then substitute $y = \cos \theta$ to eliminate θ just leaving an equation connecting x and y.

For part (b) we can solve the equation of the line with the parametric equation of the curve. Once the values of θ have been found, we can substitute them back into the parametric equations to find the x and y-coordinates.

For part (c) we need to find the expression for the gradient of the curve and then substitute the coordinates of P and Q to find the gradient at each of these two points. We can then use the gradients with the points to find the equations of each tangent. The two tangent equations can be solved simultaneously to find the coordinates of the point of intersection.

The solution

(a) $\cos 2\theta = 2 \cos^2 \theta - 1$

So, $x = 2 \cos^2 \theta - 1$ and since $y = \cos \theta$, we have:

$x = 2y^2 - 1$

$2y^2 = x + 1$

> It is useful to remember this trig identity as it is not included in the formula booklet.

(b) $x - y + 1 = 0$ and $x = \cos 2\theta, y = \cos \theta$.

$\cos 2\theta - \cos \theta + 1 = 0$

$2\cos^2 \theta - 1 - \cos \theta + 1 = 0$

$2\cos^2 \theta - \cos \theta = 0$

$\cos \theta (2 \cos \theta - 1) = 0$

$$\cos \theta = \frac{1}{2}, 0$$

$$\theta = \frac{\pi}{3}, \frac{\pi}{2}$$

For P, $\theta = \frac{\pi}{3}$ so $x = \cos 2\theta = \cos \frac{2\pi}{3} = -\frac{1}{2}$ and $y = \cos \theta = \cos \frac{\pi}{3} = \frac{1}{2}$

Hence P is $\left(-\frac{1}{2}, \frac{1}{2} \right)$

For Q, $\theta = \frac{\pi}{2}$ so $x = \cos 2\theta = \cos \pi = -1$ and $y = \cos \theta = \cos \frac{\pi}{2} = 0$

Hence Q is $(-1, 0)$

(c) $x = \cos 2\theta, y = \cos \theta$

$\frac{dx}{d\theta} = -2 \sin 2\theta$ and $\frac{dy}{d\theta} = -\sin \theta$

$\frac{dy}{dx} = \frac{dy}{d\theta} \times \frac{d\theta}{dx} = -\sin \theta \times \frac{1}{-2 \sin 2\theta} = \frac{-\sin \theta}{-4 \sin \theta \cos \theta} = \frac{1}{4 \cos \theta}$

Gradient of tangent at $P = \dfrac{1}{4 \cos \frac{\pi}{3}} = \dfrac{1}{2}$

Coordinate geometry in the (x, y) plane

> The formula for a straight line is used here.

Equation of tangent at P is $y - \dfrac{1}{2} = \dfrac{1}{2}\left(x + \dfrac{1}{2}\right)$

$$2y - 1 = x + \dfrac{1}{2}$$

$$2y = x + \dfrac{3}{2}$$

$$4y = 2x + 3$$

Gradient of tangent at $Q = \dfrac{1}{4\cos\frac{\pi}{2}}$ which is undefined as $\cos\dfrac{\pi}{2} = 0$

Equation of tangent at Q is $x = -1$

Solving the two equations of the tangents simultaneously, we obtain the point of intersection is $\left(-1, \dfrac{1}{4}\right)$.

Exam practice

1. A line has the parametric equations $x = 5t + 1$ and $y = 1 - 2t$.

   Show that the cartesian equation of the line is $y = \dfrac{7 - 2x}{5}$. [3]

2. Given that $3y^2 + 6xy - 3x^2 = 4$, show that $\dfrac{dy}{dx} = \dfrac{x - y}{x + y}$. [3]

3. A curve C has parametric equations $x = 5t^2$ and $y = 2t^4$.
   (a) Find the Cartesian equation of curve C. [3]
   (b) Point P on the curve has parameter p.
       (i) Find the gradient of the curve at point P. [2]
       (ii) Find the equation of the tangent to the curve at P. [2]

4. A curve C is given by the parametric equations:
   $$x = 2\cos 3t \qquad y = 2\sin 3t$$
   A point P with parameter p lies on curve C.
   (a) Show that the equation of the tangent to the curve at point P is
       $$y\sin 3p + x\cos 3p - 2 = 0.$$ [5]
   (b) The tangent to the curve at P meets the x-axis at point A.
       If $p = \dfrac{\pi}{3}$, find the coordinates of point A. [3]

5. A curve C has parametric equations $x = \sin\theta, y = \cos 2\theta$.
   (a) The equation of the tangent to the curve C at the point P
       where $\theta = \dfrac{\pi}{4}$ is $y = mx + c$. Find the exact values of m and c. [6]
   (b) Find the coordinates of the points of intersection of the
       curve C and the straight line
       $$x + y = 1.$$ [5]

   (WJEC Unit 3 June 2019 q 6)

7 Integration

Prior knowledge

You will need to make sure you fully understand the following from your AS studies:

- Indefinite integration as the reverse process of differentiation.

- Evaluation of definite integrals.

Quick revision

Indefinite integration is the reverse process to differentiation. When integrating indefinitely you must remember to include the constant of integration.

$$\int x^n \, dx = \frac{x^{n+1}}{n+1} + c \quad \text{(provided } n \neq -1\text{)}$$

Definite integration is integration where you have limits. The answer to a definite integral will usually be a numerical value.

Integration of x^n $(n \neq -1)$, e^{kx}, $\dfrac{1}{x}$, $\sin kx$, $\cos kx$

$$\int x^n\, dx = \frac{x^{n+1}}{n+1} + c \quad (n \neq -1)$$

$$\int e^{kx}\, dx = \frac{e^{kx}}{k} + c$$

$$\int \frac{1}{x}\, dx = \ln|x| + c$$

$$\int \sin kx\, dx = -\frac{1}{k}\cos kx + c$$

$$\int \cos kx\, dx = \frac{1}{k}\sin kx + c$$

All these formulae must be remembered as they are not included in the formula booklet.

Integration of $(ax + b)^n$ $(n \neq -1)$, e^{ax+b}, $\sin(ax + b)$, $\cos(ax + b)$

$$\int (ax + b)^n\, dx = \frac{(ax + b)^{n+1}}{(n+1)a} + c \quad (n \neq -1)$$

$$\int e^{ax+b}\, dx = \frac{e^{ax+b}}{a} + c$$

$$\int \frac{1}{ax + b}\, dx = \frac{1}{a}\ln|ax + b| + c$$

$$\int \sin(ax + b)\, dx = -\frac{\cos(ax + b)}{a} + c$$

$$\int \cos(ax + b)\, dx = \frac{\sin(ax + b)}{a} + c$$

You will be required to remember these results as they are not given in the formula booklet.

Using definite integration to find the area between two curves

To find the shaded area between the curves $y = f(x)$ and $y = g(x)$ as shown:

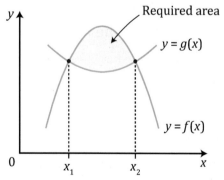

1   Find the x-coordinates of the points of intersection by solving the equations of the curves simultaneously.

2   Integrate each curve equation definitely using x_1 and x_2 as limits.

3   Subtract the two values of the integrals to give the shaded area.

Integration by substitution

An integral of the type $\int f(x)\,dx$ is converted into the integral $\int f(x)\frac{dx}{du}\,du$, where x is replaced by a given substitution. In the case of definite integrals, the x limits are converted into u limits by means of the given substitution.

Integration by parts

Integration by parts is used when there is a product to integrate (e.g. $\int x \sin x\,dx$).

The formula for integration by parts is included in the formula booklet and is:

$$\int u\frac{dv}{dx}\,dx = uv - \int v\frac{du}{dx}\,dx$$

Two rules:

1 If one of the functions is a polynomial and the other function is easily integrated, let u = polynomial so $\frac{dv}{dx}$ = other function.

2 If one of the functions is a polynomial and the other function is not easily integrated, let $\frac{dv}{dx}$ = polynomial in x and u = other function.

Integration using partial fractions

Single fractions such as $\frac{5x + 3}{(x + 3)(x + 1)}$ can be converted into partial fractions ($\frac{6}{x + 3} - \frac{1}{x + 1}$ in this case) so that each of the resulting partial fractions can be integrated. In many cases the answer to these questions involves the use of ln.

Analytical solution of first order differential equations with separable variables

Equations of the type: $\frac{dy}{dx} = f(x)\,g(y)$

can be solved by separating the variables and integrating both sides of the resulting equation, i.e.

$$\int \frac{1}{g(y)}\,dy = \int f(x)\,dx$$

Looking at exam questions

1 Find

(a) $\int \dfrac{3}{1-4x}\,dx$ [2]

(b) $\int 4e^{-3x}\,dx$ [2]

(c) $\int (6x^2 + 1)^3\,dx$ [3]

Thinking about the question

These are all integrations without limits, so a constant of integration needs to be included.

Starting the solution

For (a) we need to recognise that if we could make the differential of the bottom of the fraction equal to the top then the integral will be the ln of the bottom. So we need −4 as the numerator so we must also divide by −4 to keep the fraction the same as that given.

For part (b) we need to recall from memory the integral of e^{-3x}. Remember that the power stays the same, but we need to divide by the number in front of the x (i.e. −3 in this case).

For part (c) we notice that this is the integral of a bracket raised to a power. This is a type of integral you can expect to see on most papers. Remember the shortcut for doing this. It is included in the solution shown here.

The solution

> Notice that it is possible to make the top the differential of the bottom by multiplying the top and bottom by −4. The integral will then become ln of the bottom.

(a) $\int \dfrac{3}{1-4x}\,dx = \int \dfrac{3(-4)}{(-4)(1-4x)}\,dx$

$\qquad\qquad\quad = -\dfrac{3}{4}\int \dfrac{-4}{1-4x}\,dx$

> Don't forget to include the constant of integration, c.

$\qquad\qquad\quad = -\dfrac{3}{4}\ln(1-4x) + c$

(b) $\int 4e^{-3x}\,dx = \dfrac{-4}{3}e^{-3x} + c$

> Note that the integral of e^{ax} is not included in the formula sheet. When integrated the power of e stays the same but we need to divide by the number in front of the x (−3 in this case).

(c) $\int (6x^2 + 1)^3\,dx = \dfrac{(6x^2 + 1)^4}{4 \times 12x} + c$

$\qquad\qquad\qquad\quad = \dfrac{(6x^2 + 1)^4}{48x} + c$

> This is the integral of a bracket raised to a power. The quick way to do this is to increase the power the bracket is raised to by one. Then divide by the new power and also the differential of the inside of the bracket.

2 (a) Express $\dfrac{2x^2 + 3x + 1}{(x + 2)(x + 3)^2}$ as partial fractions. [4]

(b) Show that $\displaystyle\int \dfrac{2x^2 + 3x + 1}{(x + 2)(x + 3)^2}\, dx = \ln \dfrac{(x + 2)^3}{(x + 3)} + \dfrac{10}{x + 3} + c$ [6]

Thinking about the question

Notice that in part (a) there is a repeated linear factor (i.e. $(x + 3)^2$) in the denominator. This means we need three partial fractions.

For part (b) we are asked to find the integral of the original expression, so we need to use the result from part (a) and integrate this.

Starting the solution

For part (a) we need to have three denominators $x + 2$, $x + 3$, $(x + 3)^2$.

For part (b) we notice that there is ln involved so we need to look for the differential of the bottom on the top of the partial fractions. We may need to adjust the numbers to make this so.

The solution

$$\dfrac{2x^2 + 3x + 1}{(x + 2)(x + 3)^2} = \dfrac{A}{x + 2} + \dfrac{B}{x + 3} + \dfrac{C}{(x + 3)^2}$$

> Multiply both sides by $(x + 2)(x + 3)^2$.

$$2x^2 + 3x + 1 = A(x + 3)^2 + B(x + 2)(x + 3) + C(x + 2)$$

Let $x = -3$ $2(-3)^2 + 3(-3) + 1 = 0 + 0 + C(-3 + 2)$

$$18 - 9 + 1 = -C$$

$$C = -10$$

> We choose x to be a value that will make one or more of the brackets zero.

Let $x = -2$ $2(-2)^2 + 3(-2) + 1 = (-2 + 3)^2 A + 0 + 0$

$$8 - 6 + 1 = A$$

$$A = 3$$

Let $x = 0$ $2(0)^2 + 3(0) + 1 = (0 + 3)^2 A + B(0 + 2)(0 + 3) + C(0 + 2)$

$$1 = 9A + 6B + 2C$$

$$1 = 9(3) + 6B + 2(-10)$$

$$1 = 27 + 6B - 20$$

$$B = -1$$

Partial fractions are $\dfrac{3}{x + 2} - \dfrac{1}{x + 3} - \dfrac{10}{(x + 3)^2}$

> Ensure you substitute the values of A, B and C in to obtain the partial fractions.

> To integrate $(x + 3)^{-2}$ increase the index by one and then divide by the new index (i.e. -1) and also divide by the differential of the inside of the bracket (i.e. 1).

> You need to recall the rules of logs to simplify part of this expression.

(b) $\int \dfrac{2x^2 + 3x + 1}{(x + 2)(x + 3)^2}\, dx = \int \left(\dfrac{3}{x + 2} - \dfrac{1}{x + 3} - \dfrac{10}{(x + 3)^2} \right) dx$

$= \int \left(\dfrac{3}{x + 2} - \dfrac{1}{x + 3} - 10\,(x + 3)^{-2} \right) dx$

$= 3 \ln (x + 2) - \ln (x + 3) - \dfrac{10}{-1}(x + 3)^{-1} + c$

$= 3 \ln (x + 2) - \ln (x + 3) + \dfrac{10}{x + 3} + c$

$= \ln (x + 2)^3 - \ln (x + 3) + \dfrac{10}{x + 3} + c$

$= \ln \dfrac{(x + 2)^3}{(x + 3)} + \dfrac{10}{x + 3} + c$

3   (a)   Find $\int (e^{2x} + 6 \sin 3x)\, dx$ [2]

(b)   Find $\int 7(x^2 + \sin x)^6\,(2x + \cos x)\, dx$ [1]

(c)   Find $\int \dfrac{1}{x^2} \ln x\, dx$ [4]

(d)   Use the substitution $u = 2 \cos x + 1$ to evaluate

$\displaystyle\int_0^{\frac{\pi}{3}} \dfrac{\sin x}{(2 \cos x + 1)^2}\, dx.$ [4]

(WJEC June 2019 Unit 3 q14)

Thinking about the question

Part (a) is a simple integration of each term. For part (b) we notice that the contents of the second bracket are the derivative of the first. We need to recall a formula for this. Part (c) is a product, so we need to use the integration by parts which is included in the formula booklet. Part (d) is integration by substitution.

Starting the solution

For (a) we differentiate each term and remember to include the constant of integration. For (b) we recall the rule $\int f'(g(x))\, g'(x)\, dx = f(g(x)) + k$ and use it. For part (c) we can look up the formula $\int u \dfrac{dv}{dx}\, dx = uv - \int v \dfrac{du}{dx}\, dx$ in the formula booklet. For part (d) we differentiate u to give $\dfrac{du}{dx}$ and then rearrange to give dx and then substitute this into the integral.

The solution

(a) $\int(e^{2x} + 6\sin 3x)\,dx = \dfrac{e^{2x}}{2} - \dfrac{6\cos 3x}{3} + c$

$$= \dfrac{e^{2x}}{2} - 2\cos 3x + c$$

(b) $\int 7(x^2 + \sin x)^6 (2x + \cos x)\,dx$

> Notice that there is only 1 mark allocated for the answer to this so you must be able to simply write the answer down.

The following formula is used for the integration.

$\int f'(g(x))\,g'(x)\,dx = f(g(x)) + k$

> Note that this formula is not included in the formula booklet.

Compare the left-hand side of this equation with the given integral

$\int 7(x^2 + \sin x)^6 (2x + \cos x)\,dx$

Here $f'(g(x)) = 7(x^2 + \sin x)^6$ and $g'(x) = 2x + \cos x$

So $\int 7(x^2 + \sin x)^6 (2x + \cos x)\,dx = 7\dfrac{(x^2 + \sin x)^7}{7} + c$

$$= (x^2 + \sin x)^7 + c$$

(c) $\int \dfrac{1}{x^2}\ln x\,dx = \int x^{-2}\ln x\,dx$

$\int u\dfrac{dv}{dx}\,dx = uv - \int v\dfrac{du}{dx}\,dx$

> The formula for integration by parts is included in the formula booklet.

Let $u = \ln x$ and $\dfrac{dv}{dx} = x^{-2}$

$\int \ln x\, x^{-2}\,dx$

$= \ln x \left(\dfrac{x^{-1}}{-1}\right) - \int\left(\dfrac{x^{-1}}{-1}\right)\left(\dfrac{1}{x}\right)dx$

$= -\dfrac{1}{x}\ln x + \int\dfrac{1}{x^2}\,dx$

$= -\dfrac{1}{x}\ln x - x^{-1} + c$

$= -\dfrac{1}{x}\ln x - \dfrac{1}{x} + c$

> We need to consider carefully which part we will let equal u. The integral of $\ln x$ is not known so we do not let $\dfrac{dv}{dx} = \ln x$.

(d) $\displaystyle\int_0^{\frac{\pi}{3}} \frac{\sin x}{(2\cos x + 1)^2}\, dx$

Let $u = 2\cos x + 1$ so

$\dfrac{du}{dx} = -2\sin x$, giving $dx = -\dfrac{du}{2\sin x}$

Integral $= \displaystyle\int \frac{\sin x}{u^2}\left(-\frac{du}{2\sin x}\right)$

Changing the limits using $u = 2\cos x + 1$.

When $x = \dfrac{\pi}{3}$, $u = 2\cos\dfrac{\pi}{3} + 1 = 2$

And when $x = 0$, $u = 2\cos 0 + 1 = 3$

Integral $= -\dfrac{1}{2}\displaystyle\int_3^2 u^{-2}\, du$

$= -\dfrac{1}{2}\left[-u^{-1}\right]_3^2$

$= -\dfrac{1}{2}\left[-\dfrac{1}{u}\right]_3^2$

$= -\dfrac{1}{2}\left[\left(-\dfrac{1}{2}\right) - \left(-\dfrac{1}{3}\right)\right]$

$= -\dfrac{1}{2} \times -\dfrac{1}{6}$

$= \dfrac{1}{12}$

---

4   The variable y satisfies the differential equation:

$$2\frac{dy}{dx} = 5 - 2y \quad \text{where } x \geq 0.$$

Given that $y = 1$ when $x = 0$, find an expression for y in terms of x.

[5]

(WJEC June 2018 unit 3 q 15)

Thinking about the question

To obtain an expression for y we need to integrate to be able to separate the $\dfrac{dy}{dx}$ term.

Starting the solution

We need to start by separating the variables and integrate. This means the terms containing y will be on the opposite side of the equals sign to those containing x.

The conditions $y = 1$ when $x = 0$ are then used to find the constant of integration which can then be substituted back into the result of the integration. Most of the time these results involve ln so to remove ln we take exponentials.

The solution

$$2\frac{dy}{dx} = 5 - 2y$$

Separating variables and integrating, we obtain:

$$\int \frac{2\,dy}{5 - 2y} = \int dx$$

$$-\ln(5 - 2y) = x + c$$

$y = 1$ when $x = 0$, so $-\ln(3) = 0 + c$, hence $c = -\ln 3$.

$$-\ln(5 - 2y) = x - \ln 3$$

$$\ln(5 - 2y) = -x + \ln 3$$

$$\ln(5 - 2y) - \ln 3 = -x$$

$$\ln\left(\frac{5 - 2y}{3}\right) = -x$$

Taking exponentials of both sides, we obtain:

$$\frac{5 - 2y}{3} = e^{-x}$$

$$5 - 2y = 3e^{-x}$$

$$y = \frac{5 - 3e^{-x}}{2}$$

> We need to include a constant of integration and we must then find an expression for this constant using the values for x and y given in the question.

> The rules of logs are used here.

---

5 Actinium is a radioactive substance which decays slowly.

Initially, 2 kg of actinium is present and the rate of decay of its mass is 64 g/year. Subsequently, t years later when the actinium has a mass x kg, the rate of decrease of mass is proportional to the value of x.

(a) Show that $\dfrac{dx}{dt} = -0.032x$ [3]

(b) Deduce that $t = \dfrac{125}{4}\ln\left(\dfrac{2}{x}\right)$.   [5]

(c) Find the value of t when half the actinium has decayed, giving your answer correct to two decimal places.   [2]

(WJEC specimen paper 2005/2006 C4 q6)

Thinking about the question

This question involves the formation of a differential equation and then by separating the variables and integrating finding a value for the constant of proportionality. When this is found, the resulting equation for t in terms of x can be found. This equation can then be used for part (c).

Starting the solution

For part (a) we form an equation from the statement, remembering to include a negative sign as the rate of change is a decrease with time. We know the value of $\frac{dx}{dt}$ when $x = 2$ and this pair of values is used to work out the value of the constant of proportionality, k.

For part (b) we separate the variables and integrate. We need a constant of integration whose value needs to be found. We know that when $t = 0$, $x = 2$ and this can be used to find c which is then substituted back into the equation. We then need to manipulate the equation until it is in the form as specified in the question.

For part (c) we use the equation obtained in part (b) with the value of x (i.e. 1 kg) to find the value of t in years.

The solution

(a) The rate of decrease of x is proportional to the value of x.

$$\frac{dx}{dt} \propto -x$$

$$\frac{dx}{dt} = -kx$$

Now $\frac{dx}{dt} = -\frac{64}{1000}$ when $x = 2$, hence $-0.064 = -k \times 2$ so $k = 0.032$

Hence $\frac{dx}{dt} = -0.032x$

(b) $\frac{dx}{dt} = -0.032x$

Separating variables and integrating, we obtain:

$$\int \frac{dx}{x} = -0.032 \int dt$$

$$\ln x = -0.032t + c$$

When $t = 0$, $x = 2$

$$\ln 2 = -0.032 \times 0 + c$$

$$c = \ln 2$$

Substituting c into $\ln x = -0.032t + c$ we obtain:

$$\ln x = -0.032t + \ln 2$$

$$\ln 2 - \ln x = 0.032t$$

$$\ln\frac{2}{x} = 0.032t$$

$$t = \frac{1}{0.032}\ln\frac{2}{x}$$

$$= \frac{1000}{32}\ln\frac{2}{x}$$

$$= \frac{125}{4}\ln\frac{2}{x}$$

Note that we don't need to take exponentials to rearrange for t.

(c) When half of the actinium has decayed, $x = 1$ kg.

$$t = \frac{125}{4}\ln\frac{2}{x}$$

$$= \frac{125}{4}\ln\frac{2}{1}$$

$$= \frac{125}{4}\ln 2$$

$$= 21.66 \text{ years (2 d.p.)}$$

Exam practice

1 Integrate:
(a) e^{6-4x} [2]
(b) $x^3 \ln x$ [5]

2 (a) Given $f(x) \equiv \dfrac{4x}{(x+1)(x-3)^2}$

Show that $\dfrac{4x}{(x+1)(x-3)^2} \equiv \dfrac{A}{x+1} + \dfrac{B}{x-3} + \dfrac{C}{(x-3)^2}$

where A, B and C are constants to be found. [4]

(b) Evaluate $\displaystyle\int_4^5 \dfrac{4x}{(x+1)(x-3)^2}\,dx$. Give your answer correct to two decimal places. [6]

3 Solve the differential equation:

$$\frac{dy}{dx} = 4x^3 y$$

given that when $x = 2$, $y = 1$. Give the answer in the form $y = f(x)$. [6]

The integral of sin x is $-\cos x$. This needs to be remembered.

We need to also divide by the derivative of the contents of the bracket (i.e. -2 here) and also add the constant of integration, c.

④ Evaluate

(a) $\displaystyle\int_0^1 xe^{-3x}\,dx$ [5]

(b) $\displaystyle\int_0^1 \sqrt{(1-x^2)}\,dx$ [5]

⑤ (a) Find each of the following, simplifying your answer wherever possible:

(i) $\displaystyle\int \sin(1-2x)\,dx$ [2]

(ii) $\displaystyle\int \frac{3}{e^{3x-1}}\,dx$ [2]

(iii) $\displaystyle\int \frac{3}{\frac{1}{2}x-1}\,dx$ [2]

(b) Evaluate $\displaystyle\int_1^5 \sqrt{(2x-1)}\,dx$ [4]

⑥ Show $\displaystyle\int_0^1 \frac{8x^2\,dx}{\sqrt{(1-x^2)}} = 2\pi$ [7]

⑦ Using a suitable substitution, find $\displaystyle\int 6x^2\sqrt{(1+x^3)}\,dx$ [5]

⑧ (a) Find $\displaystyle\int \frac{4x^3+3x^2}{x^4+x^3}\,dx$ [3]

(b) Find $\displaystyle\int \frac{\cos x}{\sin^3 x}\,dx$ [5]

⑨ Find the value of $\displaystyle\int_4^5 \frac{x+1}{(x-1)(x-2)(x-3)}\,dx$ [7]

⑩ Part of the surface of a small lake is covered by green algae. The area of the lake covered by the algae at time t years is $A\,\text{m}^2$. The rate of increase of A is directly proportional to $\sqrt{A}$.

(a) Write down a differential equation satisfied by A. [1]

(b) The area of the lake covered by the algae at time $t=3$ is $64\,\text{m}^2$ and the area covered at time $t=5.5$ is $196\,\text{m}^2$. Find an expression for A in terms of t. [6]

(WJEC June 2013 C4q)

⑪ Wildflowers grow on the grass verge by the side of a motorway. The area populated by wildflowers at time t years is $A\,\text{m}^2$. The rate of increase of A is directly proportional to A.

(a) Write down a differential equation that is satisfied by A. [1]

(b) At time $t=0$, the area populated by wildflowers is $0.2\,\text{m}^2$. One year later, the area has increased to $1.48\,\text{m}^2$. Find an expression for A in terms of t in the form pq^t, where p and q are rational numbers to be determined. [7]

(WJEC June 2019 Unit 3 q13)

⑫ Find the value of $\displaystyle\int_0^{\frac{\pi}{3}} 2\cos\left(3x+\frac{\pi}{3}\right)dx$. [6]

8 Numerical methods

Prior knowledge

You will need to make sure you fully understand the following from your AS studies:

- Sketching curves of functions.

- Interpreting algebraic solutions of equations graphically.

- Using intersection points of graphs of curves to solve equations.

Quick revision

Location of roots of $f(x) = 0$, considering changes of sign of $f(x)$

If $f(x)$ can take any value between a and b, then if there is a change of sign between $f(a)$ and $f(b)$, then a root of $f(x)$ lies between a and b.

Newton–Raphson iteration

Newton–Raphson iteration for solving $f(x) = 0$

$$x_{n+1} = x_n - \frac{f(x_n)}{f'(x_n)}$$

The Trapezium rule for estimating the area under a curve or the integral of a function

The Trapezium rule can be used for estimating areas or working out definite integrals of functions where the function is too difficult to integrate.

$$\int_a^b y\,dx \approx \frac{1}{2}h\{(y_0 + y_n) + 2(y_1 + y_2 + \ldots + y_{n-1})\} \quad \text{where } h = \frac{b-a}{n}$$

Looking at exam questions

1 By drawing suitable graphs, show that $x - 1 = \cos x$ has only one root. Starting with $x_0 = 1$, use the Newton–Raphson method to find the value of this root correct to two decimal places.   [6]

(WJEC June 2018 Unit 3 q17)

Thinking about the question

For the first part we need to draw the two graphs and show there is only one point of intersection. For the second part we need to use the Newton–Raphson formula, which is included in the formula booklet.

Starting the solution

The graph $y = x - 1$ is a straight line, intersecting the y-axis at -1 and with a gradient of 1. We need to draw both $y = x - 1$ and $y = \cos x$ on the same set of axes to show the one point of intersection. Note that as $x_0 = 1$ we must work in radians, so the x-axis needs to have units in terms of π. For the next part we need to look up the Newton–Raphson formula in the formula booklet. As we are finding the x-coordinate of the point of intersection we can equate the y-values like this: $x - 1 = \cos x$.
We then express this as a function of x (i.e. $f(x) = x - 1 - \cos x$).
We then need to differentiate to find $f'(x)$. We then use the calculation with $x_0 = 1$ until we reach a value where the number rounded to 2 d.p. remains constant.

The solution

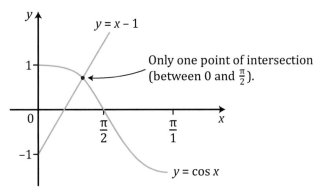

Only one point of intersection (between 0 and $\frac{\pi}{2}$).

The iteration equation we use to solve $f(x) = 0$ is:

$$x_{n+1} = x_n - \frac{f(x_n)}{f'(x_n)}$$

Note that $f'(x_n)$ is the derivative of the original function.

Now $x - 1 = \cos x$

$$f(x) = x - 1 - \cos x$$

$$f'(x) = 1 + \sin x$$

$$x_{n+1} = x_n - \frac{f(x_n)}{f'(x_n)}$$

This Newton–Raphson formula is included in the formula booklet.

$$x_0 = 1$$

$$x_1 = x_0 - \frac{f(x_0)}{f'(x_0)}$$

Note that the angles are in radians.

$$= x_0 - \frac{x_0 - 1 - \cos x_0}{1 + \sin x_0}$$

$$= 1 - \frac{1 - 1 - \cos 1}{1 + \sin 1}$$

All occurrences of x_0 are replaced by the starting value (i.e. 1).

$$= 1.293408$$

$$x_2 = x_1 - \frac{f(x_1)}{f'(x_1)}$$

$$= 1.293408 - \frac{f(1.293408)}{f'(1.293408)}$$

$$= 1.283436$$

In a similar way, x_3, x_4 are calculated.

$x_3 = 1.283429$

$x_4 = 1.283429$

Root is 1.28 (correct to 2 d.p.)

Using a calculator to perform the iteration

These instructions refer to the CASIO fx-991EX CLASSWIZ calculator.

Set the Angle Unit to Radian

Type in **1 =**   (this is the value of x_0 you are using)

Now we type in the equation with x_0 replaced by **Ans** like this

$$\text{Ans} - \frac{\text{Ans} - 1 - \cos{(\text{Ans})}}{1 + \sin{(\text{Ans})}}$$

> Note that the original expression was
> $$x_0 - \frac{x_0 - 1 - \cos x_0}{1 + \sin x_0}$$

> This is the value of x_1.

Press **=** and the following answer is displayed:

1.293407993

Press **=** again and the following value of x_2 is displayed:

1.283435773

Press **=** again and the following value of x_3 is displayed:

1.283428742

Press **=** again and the following value of x_4 is displayed:

1.283428742

> You can see the value to 2 d.p. stays constant. In this case the number is constant to more decimal places. We can now give the value of the root correct to 2 decimal places.

Root is 1.28 (correct to 2 d.p.)

---

2   A chord AB subtends an angle θ radians at the centre of a circle. The chord divides the circle into two segments whose areas are in the ratio $1 : 2$.

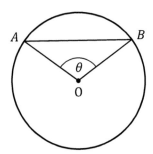

(a)   Show that $\sin\theta = \theta - \dfrac{2\pi}{3}$.   [4]

(b) (i) Show that θ lies between 2.6 and 2.7.

    (ii) Starting with $\theta_0 = 2.6$, use the Newton–Raphson method to find the value of θ correct to three decimal places. [6]

(WJEC June 2019 Unit 3 q12)

Thinking about the question

For part (a) we note that there are two segments: a large and a small one. If we find the area of the small segment, then we can find the area of the large segment by subtracting the area of the small segment from the area of the circle. We should then be able to use the ratio of the areas to show the given equation.

For part (b)(i) we have to express the given equation as a function of θ and show that when both values are substituted into the function in turn, there is a change in sign. This will show that θ lies between 2.6 and 2.7 radians.

For part (b)(ii) we obtain the Newton–Raphson formula from the formula booklet. As part of the formula, we will need to find $f'(\theta)$.

Starting the solution

To find the area of the minor segment we find the area of the sector using the formula $\frac{1}{2}r^2\theta$ and then subtract the area of the triangle using $\frac{1}{2}ab\sin C$. We must remember both these formulas and that the angle unit on our calculator must be set to Radian. To find the area of the major segment we take the area of the minor segment just found and subtract it from the area of the whole circle found using πr^2. We can then form an equation using the ratio of the two areas.

For (b)(i) we need to form an equation for $f(\theta)$ and then substitute 2.6 and 2.7 in turn and if there is a sign change then the solution lies between these two values.

For part (b)(ii) we use the Newton–Raphson formula, obtained from the formula booklet to find a value for θ. We keep repeating the iteration until the value is constant to the third decimal place.

The solution

(a) Area of the sector bounded by the angle $\theta = \frac{1}{2}r^2\theta$

Area of triangle $OAB = \frac{1}{2}r^2\sin\theta$

> This formula is not included in the formula booklet.

> To find the area of the triangle we must recall the formula
> Area $= \frac{1}{2}ab\sin C$

Area of small segment = area of sector – area of triangle

$$= \frac{1}{2}r^2\theta - \frac{1}{2}r^2\sin\theta$$

$$= \frac{1}{2}r^2(\theta - \sin\theta)$$

Area of large segment = area of circle – area of small segment

$$= \pi r^2 - \frac{1}{2}r^2(\theta - \sin\theta)$$

Area of large segment = 2 × area of small segment

$$\pi r^2 - \frac{1}{2}r^2(\theta - \sin\theta) = 2\left[\frac{1}{2}r^2\theta - \frac{1}{2}r^2\sin\theta\right]$$

$$\pi r^2 - \frac{1}{2}r^2\theta + \frac{1}{2}r^2\sin\theta = r^2\theta - r^2\sin\theta$$

Dividing through by r^2

$$\pi - \frac{1}{2}\theta + \frac{1}{2}\sin\theta = \theta - \sin\theta$$

$$\frac{3}{2}\sin\theta = \frac{3}{2}\theta - \pi$$

Dividing both sides by $\frac{3}{2}$, we obtain:

$$\sin\theta = \theta - \frac{2\pi}{3}$$

> Remember to set your calculator for radian angles.

(b) (i) $f(\theta) = \sin\theta - \theta + \frac{2\pi}{3}$

When $\theta = 2.6, f(2.6) = \sin 2.6 - 2.6 + \frac{2\pi}{3} = 0.0099$

When $\theta = 2.7, f(2.7) = \sin 2.7 - 2.7 + \frac{2\pi}{3} = -0.18$

> The zero value for θ must lie between these two values.

As there is a sign change between these two values, θ must lie between these two values.

$$\theta_{n+1} = \theta_n - \frac{f(\theta_n)}{f'(\theta_n)}$$

$$f'(\theta) = \cos\theta - 1$$

> The Newton–Raphson formula is obtained from the formula booklet.

$$\theta_1 = \theta_0 - \frac{f(\theta_0)}{f'(\theta_0)}$$

$$= \theta_0 - \frac{\sin \theta_0 - \theta_0 + \frac{2\pi}{3}}{\cos \theta_0 - 1}$$

$$= 2.6 - \frac{\sin 2.6 - 2.6 + \frac{2\pi}{3}}{\cos 2.6 - 1}$$

$$= 2.6053296$$

$$\theta_2 = 2.605325675$$

$$\theta = 2.605 \text{ rad (correct to 3d.p.)}$$

To perform the calculation using a calculator:

First set you calculator to use radians for angles.

Type 2.6 =

Type in the equation replacing θ_0 with Ans like this

$$\text{Ans} - \frac{\sin(\text{Ans}) - \text{Ans} + \frac{2\pi}{3}}{\cos(\text{Ans}) - 1}$$

Keep pressing = until you obtain a constant value correct to 3 d.p.

---

3 (a) Use the Trapezium rule with five ordinates to find an approximate value for the integral:

$$\int_5^7 \ln(1 + x^2)\, dx$$

Show your working and give your answer correct to one decimal place. [4]

(b) The graph of $y = \ln(1 + x^2)$ is shown below:

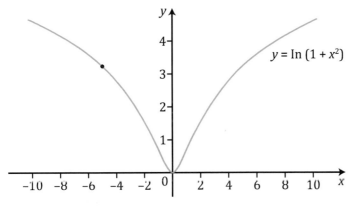

Explain, by giving a reason, whether the area you calculated in part (a) will be larger or smaller than the accurate area. [1]

Thinking about the question

This is a question about finding the area under a curve between two limits using the Trapezium rule.

For part (b) we need to think about whether the curve is above or below the lines that form the tops of the trapeziums.

Starting the solution

The formula for the Trapezium rule is obtained from the formula booklet. We first work out the width of a strip using $h = \dfrac{b-a}{n}$. We then substitute the lower limit into the equation for the curve to find y_0. We then increase x by h and substitute this value into the equation of the curve to find y_1. We keep doing this until we reach the upper limit (in this case 7). This last value gives y_n. We then substitute these y-values into the formula and give the answer to one decimal place.

For part (b) we can see that the curve is above the tops of the trapeziums used to calculate the area meaning the area calculated will be an underestimate of the actual area.

The solution

(a) $h = \dfrac{b-a}{n} = \dfrac{7-5}{4} = 0.5$

When $x = 5, y_0 = \ln(1 + 25) = 3.2581$

$x = 5.5, y_1 = \ln(1 + 5.5^2) = 3.4420$

$x = 6.0, y_2 = \ln(1 + 6^2) = 3.6109$

$x = 6.5, y_3 = \ln(1 + 6.5^2) = 3.7670$

$x = 7.0, y_n = \ln(1 + 7^2) = 3.9120$

$$\int_a^b y\, dx \approx \frac{1}{2}h\{(y_0 + y_n) + 2(y_1 + y_2 + \ldots + y_{n-1})\}$$

$$\int_5^7 \ln(1 + x^2)\, dx \approx \frac{1}{2} \times 0.5\{(3.2581 + 3.9120) + 2(3.4420 + 3.6109 + 3.7670)\}$$

$$\approx 7.202475$$

$$\approx 7.2\ (1\ \text{d.p.})$$

(b) The curve lies above the trapeziums, so there will be an area not included when the Trapezium rule is used. Hence the area calculated will be less than the actual area.

> *h* is the width of each strip.

> The formula for the Trapezium rule is obtained from the formula booklet.

▶▶▶▶▶ Exam practice

1 (a) Using the same set of axes, sketch the graph of $y = e^{-x}$ and $y = x$ and show that the equation $e^{-x} = x$ only has one root. [4]

(b) Use the Newton–Raphson method with $x_0 = 0.6$ to find the solution of the equation $e^{-x} = x$ correct to 3 decimal places. [6]

2 By drawing suitable graphs, show that $\sin x + x - 1 = 0$ has only one root. Starting with $x_0 = 0.5$, use the Newton–Raphson method to find the value of this root correct to four decimal places. [8]

3 (a) (i) Find $\displaystyle\int_0^a (e^{2x} - 1)\, dx$.

 (ii) Given that $\displaystyle\int_0^a (e^{2x} - 1)\, dx = \frac{1}{2}(9 - a)$

 show that $e^{2a} - a - 10 = 0$. [4]

 (b) Show that the equation $e^{2a} - a - 10 = 0$ has a root α between 1 and 2.

 The recurrence relation $a_{n+1} = \dfrac{1}{2}\ln(a_n + 10)$ with $a_0 = 1.2$

 can be used to find α. Find and record the values of a_1, a_2, a_3, a_4. Write down the value of a_4 correct to five decimal places and prove that this value is the value of α correct to five decimal places. [7]

 (WJEC June 2013 C3 q4)

4 (a) Sketch the graphs of $y = x^2$ and $y = \cos x$ on the same axes for $0 \le x \le \dfrac{\pi}{2}$ and show that there is one root for the equation $x^2 - \cos x = 0$ in this range. [3]

 (b) Show that this root, a, lies in the interval $0.8 < a < 0.9$. [2]

 (c) A student uses the iteration formula $x_{n+1} = \sqrt{\cos x_n}$ to find an approximation for the root a. They start with $x_0 = 0.8$. Determine whether or not this iteration formula can be used to find an approximate value of a, giving a reason for your answer. [4]

5 (a) Use the Trapezium rule with five ordinates to find an approximate value for the integral:

$$\int_0^4 \frac{1}{x^2 + 2}\, dx$$

 Show your working and give your answer correct to one decimal place. [6]

 (b) The graph of $y = \dfrac{1}{x^2 + 2}$ shown below.

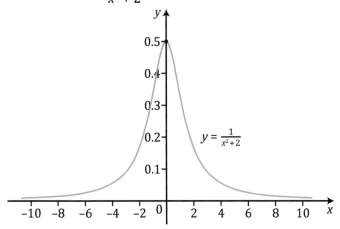

 Explain, by giving a reason, whether the area you calculated in part (a) will be larger or smaller than the accurate area. [2]

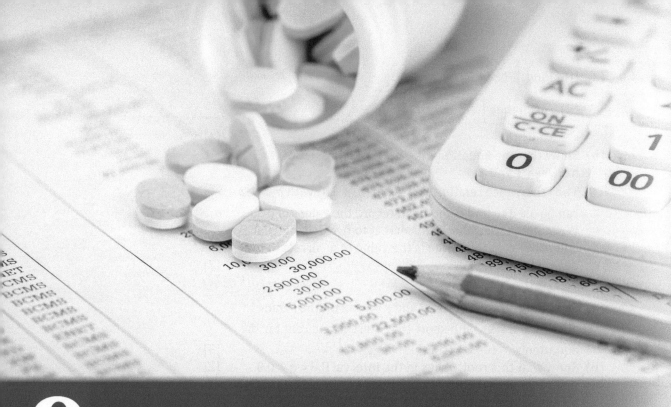

9 Probability

Prior knowledge

You will need to make sure you understand the following from your AS studies:

- Venn diagrams.
- The addition law for mutually exclusive events.
- The generalised addition law.
- The multiplication law for independent events.

Quick revision

The multiplication law for independent events

If events A and B are independent:

$P(A \cap B) = P(A) \times P(B)$

The multiplication law for dependent events

If events A and B are dependent:

$P(A \cap B) = P(A) \times P(B \mid A)$

$P(A \cap B) = P(B) \times P(A \mid B)$

> The As and Bs can be swapped around in this formula and you can also swap A for A' and B for B' for all combinations.

The generalised addition law

$P(A \cap B) = P(A) + P(B) - P(A \cup B)$

The generalised addition law can be used for dependent or independent events.

> This formula is included in the formula booklet.

The conditional probability formula

$P(A \mid B)$ means the probability of A given that B has occurred.

$P(A \mid B) = \dfrac{P(A \cap B)}{P(B)}$

> This formula is included in the formula booklet.

> The As and Bs can be swapped around in this formula and you can also swap A for A' and B for B' for all combinations. For example, you can have $P(B \mid A) = \dfrac{P(A \cap B)}{P(A)}$

Looking at exam questions

1   A bag contains 2 red balls, 3 green balls and 5 black balls.

Three balls are selected at random from the bag without replacement.

Find the probability that:

(a)   the three balls are of different colours   [4]

(b)   the three balls are the same colour.   [2]

Thinking about the question

This is a question concerning conditional probability as the probability of selecting the next ball depends on what colour has been selected before. We might decide to draw a tree diagram would take time to do and it would become quite large, so we need to think about an easier way.

Starting the solution

For part (a), there are 6 ways of picking three different coloured balls and the probability of picking each of these combinations is the same. We can simply work out the probability of one of these combinations and multiply it by 6 to give the answer.

For part (b) we find the probability of three greens and three blacks and add the two probabilities together. We note that the probability of 3 reds = 0 as there are only two reds.

The solution

(a) $P(3 \text{ different coloured balls}) = 6 \times \dfrac{2}{10} \times \dfrac{3}{9} \times \dfrac{5}{8} = \dfrac{1}{4}$

(b) $P(3 \text{ the same colour}) = P(3 \text{ green balls}) + P(3 \text{ black balls})$

$$= \dfrac{3}{10} \times \dfrac{2}{9} \times \dfrac{1}{8} + \dfrac{5}{10} \times \dfrac{4}{9} \times \dfrac{3}{8}$$

$$= \dfrac{1}{120} + \dfrac{1}{12}$$

$$= \dfrac{11}{120}$$

---

2   Val buys electrical components from one of 3 suppliers A, B, C, in the ratio $2 : 1 : 7$. The probability that the component is faulty is 0.33 for A, 0.45 for B and 0.05 for C. Val selects a component at random.

(a)   Find the probability that the component works.   [3]

(b)   Given that the component works, find the probability that Val bought the component from supplier B.   [2]

(WJEC June 2109 Unit 4 q1)

Thinking about the question

This is a question about probability. This question concerns the probability of selecting a particular supplier and then the probability of a particular component working.

We need to think about how to work out the probability of picking a particular supplier given the ratios.

Starting the solution

To find the probability of choosing a particular supplier from the ratio, we add the ratios together (i.e. 2 + 1 + 7 =10) and then divide each ratio by the total. So, the probability of choosing supplier A is $\frac{2}{10}$ etc.

For part (a) we can work out the probability that a component works by using 1 – Probability that the component is faulty. We can then use the AND law for working out the probability of using a particular supplier and then having a component that works. There will be three such calculations and these probabilities need to be added together.

For part (b) this is a question on conditional probability so we can use the conditional probability formula obtained from the formula sheet.

The solution

(a) $2 + 1 + 7 = 10$

$$P(A) = \frac{2}{10}, P(B) = \frac{1}{10}, P(C) = \frac{7}{10}$$

P(Component from A works) $= 1 - 0.33 = 0.67$

P(Component from B works) $= 1 - 0.45 = 0.55$

P(Component from C works) $= 1 - 0.05 = 0.95$

P(Component is from A and works) $= \frac{2}{10} \times 0.67 = 0.134$

P(Component is from B and works) $= \frac{1}{10} \times 0.55 = 0.055$

P(Component is from C and works) $= \frac{7}{10} \times 0.95 = 0.665$

P(Component works) $= 0.134 + 0.055 + 0.665 = 0.854$

(b) We need to find the probability that the component is from supplier B given that it works.

The following conditional probability formula is looked up from the formula booklet.

$$P(A \cap B) = P(A)\, P(B \mid A)$$

$$P(B \mid A) = \frac{P(A \cap B)}{P(A)} = \frac{0.055}{0.854} = 0.0644$$

> The total of the ratios is found so the probability of choosing a particular supplier can be found.

3 Four children are playing a game in order to win a calculator. They take turns, starting with Alex, followed by Ben, then Caroline, then Danielle, rolling a fair six-sided dice until someone obtains a 6. This player then wins a calculator.

(a) Find the probability that:

(i) Danielle wins the calculator on her first turn. [1]

(ii) Ben wins the calculator on his first or second turn. [3]

(iii) Caroline rolls the dice exactly twice. [3]

(b) Show that the probability that Alex wins the calculator is $\frac{216}{671}$. [3]

(WJEC June 2109 Unit 4 q2)

Thinking about the question

For part (a) we need to consider what needs to happen before we get to the required event. For example, for part (a) (i) Alex, Ben and Caroline would need to lose for Danielle to win on her first turn.

For part (b) it first looks like something is missing as we do not know how many times the game is played. Theoretically it could be played forever. We then think this question is about a series where we could find the sum to infinity.

Starting the solution

For part (a)(i) we can list the events and multiply the probabilities.

For part (a)(ii) we need to work out the probability of Ben winning on his first turn and then work out the probability of winning on his second turn and add them together.

For part (a)(iii) we need to consider for her to roll twice she would have to lose on her first turn.

For part (b) we can work out the probability of Alex winning on his first few turns. As we don't know how many times he needs to throw, this will be a question on summing to infinity so we need to determine the first term and the common ratio and put them into the formula for the sum to infinity.

The solution

(a) (i) P(D wins on 1st turn) = P(A lose) × P(B lose) × P(C lose)
$$\times \text{ P(D win)}$$
$$= \frac{5}{6} \times \frac{5}{6} \times \frac{5}{6} \times \frac{1}{6}$$
$$= \frac{125}{1296}$$

(ii) P(B wins on 1st turn) = $\dfrac{5}{6} \times \dfrac{1}{6} = \dfrac{5}{36}$

This is the probability that A loses and B wins on his first turn.

Here A, B, C and D lose on their first turn and A loses on his second turn and then B wins.

P(B wins on 2nd turn) = $\left(\dfrac{5}{6}\right)^5 \times \left(\dfrac{1}{6}\right)$

$= \dfrac{3125}{46656}$

P(B wins on 1st or 2nd turn) = $\dfrac{5}{36} + \dfrac{3125}{46656} = \dfrac{9605}{46656}$

(iii) P(Caroline rolls twice) = P(Caroline wins on 2nd roll) +
P(Caroline doesn't win on 2nd roll but D wins on 2nd roll)
+P(Caroline doesn't win on 2nd roll but A wins on 2nd roll
+ P(Caroline doesn't win on 2nd roll but B wins on 2nd roll)

P(Caroline rolls twice) = $\left(\dfrac{5}{6}\right)^6 \times \left(\dfrac{1}{6}\right) + \left(\dfrac{5}{6}\right)^7 \times \left(\dfrac{1}{6}\right) + \left(\dfrac{5}{6}\right)^8 \times$
$\left(\dfrac{1}{6}\right) + \left(\dfrac{5}{6}\right)^9 \times \left(\dfrac{1}{6}\right)$

$= 0.1734$

(b) P(Alex wins) = $\left(\dfrac{1}{6}\right) + \left(\dfrac{5}{6}\right)^4 \times \dfrac{1}{6} + \left(\dfrac{5}{6}\right)^8 \times \dfrac{1}{6} + \left(\dfrac{5}{6}\right)^{12} \times \dfrac{1}{6} + \cdots$

This is a geometric series with a common ratio of $\left(\dfrac{5}{6}\right)^4$.

Using the formula for the sum to infinity we have

P(Alex wins) = $\dfrac{a}{1-r} = \dfrac{\frac{1}{6}}{1-\left(\frac{5}{6}\right)^4} = \dfrac{216}{671}$

---

4 Two football matches are played between two teams Cowbridge and Bridgend. The captain of Cowbridge reckons the probability of winning the home match is 0.6 and the probability of the away match is 0.3 and the probability of winning both the home and away matches is 0.2.

(a) Show that the probability Cowbridge does not win either match is 0.3. [2]

(b) Find the probability that team Cowbridge wins exactly one match. [2]

(c) If team Cowbridge does not win the home match, find the probability that they win the away match. [3]

Thinking about the question

This is a question about the probability of a particular team winning, drawing or losing two football matches. We need to think about the probability of just Cowbridge as all the questions refer to the probability of Cowbridge. We can represent the probabilities using a Venn diagram and use H for the home match and A for the away match.

Starting the solution

For part (a) we draw the Venn diagram. We start with the intersect representing the probability of a winning both matches and mark 0.2. We then can work outwards remembering to subtract 0.2 from each of the figures. So, the probability of only winning the home match is 0.6 – 0.2 = 0.4 and the probability of winning the away match is 0.3 – 0.2 = 0.1. We can then find the probability of not winning either match by adding all the probabilities on the diagram and subtracting the result from one.

For part (b) we just add the sections representing winning one match together.

For part (c) we use the conditional probability formula $P(A \mid B) = \dfrac{P(A \cap B)}{P(B)}$ which we obtain from the formula booklet.

The solution

(a)

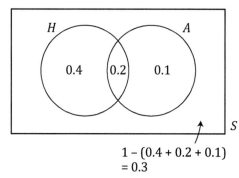

$$1 - (0.4 + 0.2 + 0.1)$$
$$= 0.3$$

P(Does not win either) = 0.3

(b)   P(Wins exactly one match) = P(wins home match only) + P(wins away match only)

$$= 0.4 + 0.1$$
$$= 0.5$$

(c) $P(A \mid H') = \dfrac{P(A \cap H)}{P(H')}$

$$= \dfrac{0.2}{(0.1 + 0.3)}$$

$$= 0.5$$

This formula is adapted from the formula for conditional probability
$P(B \mid A) = \dfrac{P(A \cap B)}{P(A)}$
obtained from the formula booklet.

Exam practice

① If $P(A) = 0.5$, $P(B) = 0.6$ and $P(A \cap B) = 0.3$, find:
  (a) $P(A \cup B)$ [2]
  (b) $P(A \cup B)'$ [1]
  (c) $P(B|A)$ [2]

② Simon has 3 types of DVDs: 5 war films, 3 cowboy films and 2 horror films. He selects 3 of the DVDs at random to watch one evening. Calculate the probability that he selects:
  (a) 1 film of each type, [3]
  (b) 3 war films, [2]
  (c) 3 films all of the same type. [3]

  (WJEC June 13 S1 q2)

③ Using the two-way table shown here, find each of the following:

	B	B'	Total
A	22	24	46
A'	28	38	66
Total	50	62	112

  (a) $P(A')$ [1]
  (b) $P(B' \cap A')$ [2]
  (c) $P(A \cup B)$ [2]

④ The events A and B are independent such that $P(A) = 0.7$ and $P(B) = 0.4$.
  (a) Find $P(A \cup B)$. [3]
  (b) Find the probability that:
    (i) exactly one of A and B will occur
    (ii) neither A nor B will occur. [6]

  (WJEC Spec 2005/2006 S1 q3)

⑤ Events A and B are such that:
  $P(A) = 0.7$, $P(B) = 0.5$, $P(A \cup B) = 0.85$.
  (a) Show that A and B are independent events. [3]
  (b) Calculate the probability of exactly one of the two events occurring. [3]

6   The two-way table shows the favourite subject between boys and girls for Maths and English in a group of 100 students.

	Boys (B)	Girls (G)	Total
Maths (M)	24	28	52
English (E)	22	26	48
Total	46	54	100

(a) Find P(M)   [1]

(b) Find P($M \cap G$)   [1]

(c) Find P($M \mid G$)   [2]

7   At a particular time it is known that 6% of the population are suffering from a certain virus. A test is used to detect the virus. When the test is given to a person who has the virus it gives a positive response with probability 0.99. When the test is given to a person who does not have the virus it gives a positive result with a probability of 0.05.

(a) Find the probability that two people selected at random, have the virus.   [2]

(b) Show that the probability of a person who does not have the virus giving a negative response is 0.893.   [3]

The test is given to a randomly selected member of the population.

(c) Find the probability that the test gives a positive response.   [3]

(d) Given that a positive response is obtained, find the probability that the person has the virus.   [2]

10 Statistical distributions

Prior knowledge

You will need to make sure you understand the following from your AS studies:

- The binomial distribution as a model.

- The Poisson distribution as a model.

- The discrete uniform distribution as a model.

- Selecting an appropriate probability distribution.

Quick revision

Continuous uniform distribution

If $X \sim U[a,b]$, then:

$$\text{Mean, } E(X) = \frac{1}{2}(a + b)$$

$$\text{Variance, } \text{Var}(X) = \frac{1}{12}(b - a)^2$$

Both of these formulae are included in the formula booklet.

$$P(c \le X \le d) = \frac{d - c}{b - a}$$

This formula is not included in the formula booklet and will need to be remembered.

Normal distribution

If $X \sim N(\mu, \sigma^2)$ the distribution is:

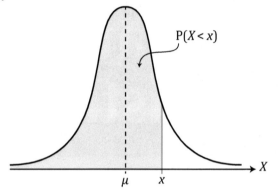

For the standard normal distribution, the distribution is adjusted to

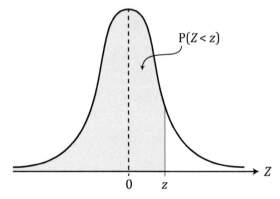

Where $z = \dfrac{x - \mu}{\sigma}$

Hints about using the normal distribution

Finding a probability – if you are finding a probability and you know the lower and upper values, the mean and standard deviation then use 'Normal CD' on your calculator.

Using the probability (or area) to find a value of x – use the area (same as probability) along with the mean, standard deviation and then use 'Inverse Normal' on your calculator.

If you need to find the mean (μ) or standard deviation (σ) or both, then you will need to use the Standard Normal Distribution N(0, 1) to find the z-value. To find the z-value using a calculator you use 'Inverse Normal' and enter the area (probability), 0 for the mean (μ) and 1 for the standard deviation (σ). You can then substitute your z-value into the formula $z = \dfrac{x - \mu}{\sigma}$ with the values for x and either μ or σ to find the required value. It is important to note that you are not using the standard normal distribution and are now using the normal distribution applicable to your question.

Use of >, ≥ or <, ≤ when using the normal distribution

$P(X > 10) = P(X \geq 10)$ and $P(X < 10) = P(X \leq 10)$ it does not matter to a question using the Normal Distribution as the probabilities are the same.

Looking at exam questions

1   A continuous random variable X is uniformly distributed over the interval [–2, 8].

    (a) Write down the mean of X.   [1]

    (b) Find $P(X \leq 3.5)$.   [2]

    (c) Find $P(1 \leq X \leq 3)$.   [2]

Thinking about the question

We need to remember that a continuous uniform distribution is a rectangular distribution and that the area under the graph between the intervals (i.e. –2 and 8) is equal to one. For part (a) we need the formula for the mean which is included in the formula booklet. For parts (b) and (c) we need to find the areas representing the range of values of X.

Starting the solution

For part (a) we obtain the formula, Mean $E(X) = \dfrac{1}{2}\left(a + b\right)$ where a is the lower value in the interval (i.e. –2) and b is the upper value (i.e. 8).

For parts (b) and (c) we can sketch the distribution noting that the maximum value on the y-axis representing $f(X)$ is $\frac{1}{b-a}$. We can then mark on the diagrams the required ranges and find the areas.

The solution

(a)   Mean $E(X) = \frac{1}{2}(a + b)$

$$= \frac{1}{2}(-2 + 8)$$

$$= 3$$

(b)

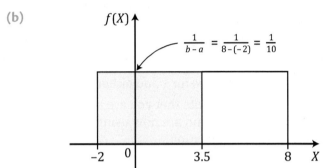

$$\frac{1}{b-a} = \frac{1}{8-(-2)} = \frac{1}{10}$$

$P(X \le 3.5)$ = area shaded

$$= \frac{1}{10} \times 5.5$$

$$= 0.55$$

(c)

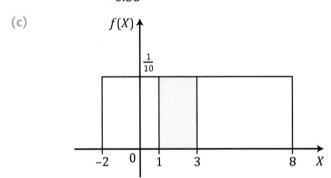

$P(1 \le X \le 3)$ = area shaded

$$= \frac{1}{10} \times 2$$

$$= 0.2$$

2 The continuous random variable X is uniformly distributed on the interval $[a, b]$ where $0 < a < b$.

(a) Let $Y = \sqrt{X}$.

Find an expression for $P(Y \leq y)$ for $\sqrt{a} \leq y \leq \sqrt{b}$. [3]

(b) Given that the mean and the variance of X are 5.5 and 3 respectively, find the values of a and b. [5]

(WJEC June 2015 S2 q7)

Thinking about the question

On reading the first sentence we need to ask why we are given the inequality $0 < a < b$. This tells us that a and b are positive values and that the value of b is greater than a.

As the random variable is X, we need to deal with X rather than $\sqrt{X}$.

Starting the solution

For part (a) we replace the Y by $\sqrt{X}$ and as we need to use X, we then need to square to remove the square root.

For part (b) we obtain the formulas for the mean and variance in terms of the parameters a and b from the formula booklet. We then create two equations, as we know the values of the mean and variance. These two equations can then be solved simultaneously.

The solution

(a) $P(Y \leq y) = P(\sqrt{X} \leq y)$

> Use the given substitution. So write $\sqrt{X}$ instead of Y.

Now we need to use X so we need to square.

$P(Y \leq y) = P(X \leq y^2)$

(b) Mean $E(X) = \dfrac{1}{2}(a + b) = 5.5$

Hence, $a + b = 11$ (1)

Variance $\text{Var}(X) = \dfrac{1}{12}(b - a)^2 = 3$

Hence, $(b - a)^2 = 36$ so $b - a = \pm 6$

> As $0 < a < b$ this means both a and b are greater than 0 and that b is greater than a. This means $b - a$ cannot be negative.

so $b - a = 6$ (2)

Adding equations (1) and (2) we have:

$2b = 17$ so $b = 8.5$

$b - a = 6$ so $8.5 - a = 6$, so $a = 2.5$

Hence $a = 2.5$ and $b = 8.5$

> We know $b - a$ cannot be minus 6 as both a and b are positive and $b > a$.

3 At a fairground, Kirsty throws n balls to try to knock coconuts off their stands. Any coconuts she knocks off are replaced before she throws again. Kirsty counts the number of coconuts she successfully knocks off their stands. On average, she knocks off a coconut with 20% of her throws.

(a) What assumptions are needed in order to model this situation with a binomial distribution? Explain whether these assumptions are reasonable. [2]

Kirsty uses a spreadsheet to produce the following diagrams, showing the probability distributions of the number of coconuts knocked off their stands for different values of n.

Distribution for $n = 6$

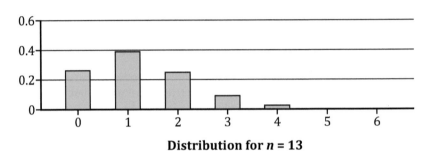

Distribution for $n = 13$

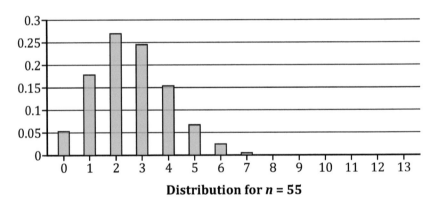

Distribution for $n = 55$

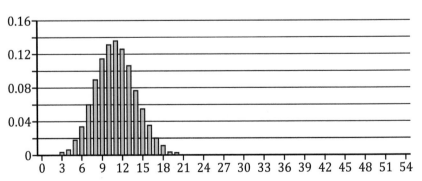

(b) Describe two ways in which the distribution changes as n increases. [2]

(WJEC June 2019 Unit 4 q3)

Thinking about the question

For part(a) we need to recall the assumptions we make when modelling using the binomial distribution and we must make our answer relevant to this modelling situation.

For part (b) we need to carefully look at the three diagrams to spot the differences between them. We initially notice that as n increases the distributions becomes more symmetrical about a central mean value. We need to think about what statistical measures can be obtained from these graphs and how these measures differ.

Starting the question

In part (a) we need to consider what might change the probability and what will change when the number of throws increases.

In part (b) in addition to the change in shape of each graph we need to consider the statistical averages (i.e. mean, mode and median) and if they could be worked out from the curve and how they might change.

The solution

(a) Each throw is independent.

The probability of 0.2 stays constant for each throw.

Reasons:

The assumptions are not reasonable as they will likely get more accurate with practice so the probability will not stay constant at 0.2.

Alternative answer:

The assumptions are not reasonable because if they failed to knock them off, they could get frustrated and the probability of knocking them off would decrease.

(b) There are lots of possible answers here including:

The distribution becomes more symmetrical.

It becomes more bell-shaped.

The mean increases.

The mode increases.

The standard deviation/variance increases.

4   Ioannis hires sunbeds and parasols on a beach. His takings per day X Euros can be modelled by a normal distribution with mean 450 Euros and standard deviation 15 Euros.

(a) Explain why he has decided to model his takings using a normal distribution. [1]

(b) Find $P(X < 425)$. [2]

(c) Find $P(X \geq 460)$. [2]

(d) Find $P(435 < X < 470)$. [2]

(e) If $P(X < x) = 0.80$, find the value of x. [2]

Thinking about the question

This question is about a normal distribution, so the random variable X is modelled as $X \sim N(\mu, \sigma^2)$. For (a) we need to recall the features of this model. For example, his takings are likely to be symmetrical about the mean value and they will tail off as the takings decrease or increase either side of the mean.

For parts (b), (c) and (d) we need to draw each distribution and work out the areas required. We will use a calculator for this.

For part (e) we need to use the 'Inverse Normal' on the calculator as we are now working back from the probability (i.e. the area) to find a value of x.

Starting the solution

For part (a) we need to mention that the data is continuous and can considered to be a random variable. We need to also mention the fact that the distribution is symmetrical about the mean value.

For part (b) we draw a normal distribution and mark on the mean. Now as the value 425 is less than the mean we draw it to the left of the mean. As we want the probability of less than this value, we shade to the left of 425. We can then see what values we need to put into the calculator. The tail has an infinitely small value, so we enter 1×10^{-99} for the lower value and for the upper value we enter 425.

For part (c) we draw a normal distribution and mark on the mean. We then mark on 460 and shade the area to the right of this. We then use the calculator with Normal CD and lower value 460 and upper value 1×10^{99}.

For part (d) we draw a normal distribution and mark on the mean and the values 435 and 470 and shade the area between. We then use Normal CD with lower value 435 and upper value 470 to work out the area (i.e. probability).

The solution

(a)   His takings can be considered as a continuous random variable and will have a symmetrical distribution about the mean value.

(b) $X \sim N(450,225)$

It is always advisable to draw a sketch of the distribution:

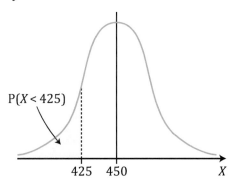

$P(X < 425)$

425   450 X

> $X \sim N(\mu, \sigma^2)$ so $X \sim N(450,225)$ tells us the mean is 450 and the variance is 225. The standard deviation $= \sqrt{\text{variance}}$ so here the standard deviation, $\sigma = \sqrt{225} = 15$.

To find $P(X < 425)$ using the CASIO Classwiz calculator, take the following steps:

Select 'Distribution' and then 'Normal CD'

Now enter the following parameters:

Lower: 1×10^{-99}

Upper: 425

σ: 15

μ: 450

> Note that the lower tail of the distribution tails off, so we use the very small value 1×10^{-99}.

The answer is displayed as P =0.04779 ...

$P(X < 425) = 0.0478$

(c)

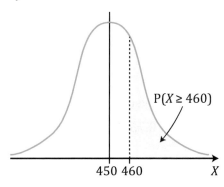

$P(X \geq 460)$

450 460 X

> It is a good idea to draw a sketch of the normal distribution. You can then see the area and hence the probability you have to find as well as identify the lower and upper values needed to calculate the probability.

10 Statistical distributions

Note that P($X \geq 460$) has the same value as P($X > 460$). So $\geq$ can be used interchangeably with $>$ as can $\leq$ with $<$.

The right tail approaches infinity so we use the very large value 1×10^{99} as the upper value.

To find P($X \geq 460$) using the CASIO Classwiz calculator, take the following steps:

Select 'Distribution' and then 'Normal CD'

Now enter the following parameters:

Lower: 460

Upper: 1×10^{99}

σ: 15

μ: 450

The answer is displayed as P = 0.2524 ...

P($X \geq 460$) = 0.252

(d) P($435 < X < 470$)

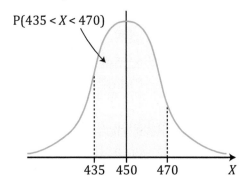

P($435 < X < 470$)

435 450   470 X

To find P($435 < X < 470$) using the CASIO Classwiz calculator, take the following steps:

Select 'Distribution' and then 'Normal CD'

Now enter the following parameters:

Lower: 435

Upper: 470

σ: 15

μ: 450

The answer is displayed as P = 0.7501 ...

P($435 < X < 470$) = 0.750

The lower and upper values use the lower and upper values in the range of values that X can take.

(e) If $P(X < x) = 0.80$, we must be looking at a value of x which is to the right of the mean.

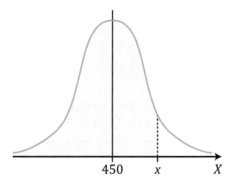

Use Inverse Normal on the calculator and enter the following parameters

Area: 0.8

σ: 15

μ: 450

The calculator gives a value xInv = 462.624.

The value of x is 462.624 Euros.

> Here you are working backwards from the area/probability to find the value of x.

---

5 A machine fills cans with X grams of baked beans where X is normally distributed with mean of 420 g and standard deviation 8 g. A can is selected at random.

(a) Find the probability that the can contains more than 430 g of beans. [2]

(b) The weight stated on the can is w grams. Find w if $P(X < w) = 0.007$. [3]

(c) The weight Y g of beans put into a can by a **different** machine is normally distributed with mean μ and standard deviation σ. If $P(Y < 400) = 0.01$ and $P(Y > 440) = 0.008$ find the values of μ and σ. [5]

Thinking about the question

For part (a) we need to draw a diagram and shade the area to find before entering the parameters into the calculator.

For (b) we need to work back from a probability and hence area to a value for w.

For (c) we are given two probabilities, and this is because there are two unknowns to find: the mean μ and the standard deviation σ.

Starting the solution

For (a) we use the calculator to find the probability to the right of the value 430. For (b) we notice that the probability is small, which means the area we need to shade on the diagram is in the lower tail. We can then mark w on the horizontal axis. We then need to use Inverse Normal Distribution on the calculator to find the value w for the given area.

For part (c) we do not know μ or σ so we cannot use the calculator directly. Instead we need to use the standard normal distribution $N(0, 1)$ and find the z-values for each of the given probabilities. We can then use the equation $z = \dfrac{x - \mu}{\sigma}$ with the pairs of values for x and z to obtain two equations in μ and σ which can then be solved simultaneously.

The solution

(a) $X \sim N(420, 8^2)$

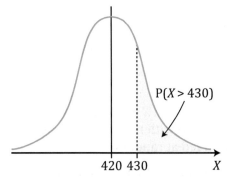

To find $P(X > 430)$ select 'Distribution' and then 'Normal CD'

Now enter the following parameters:

Lower: 430

Upper: 1×10^{99}

σ: 8

μ: 420

The answer is displayed as P = 0.1056 ...

Hence probability = 0.1056

(b)

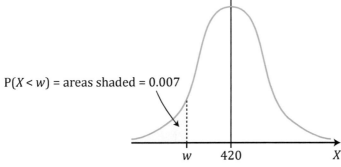

$P(X < w)$ = areas shaded = 0.007

w 420 X

Use the 'Inverse Normal' on the calculator with $\mu = 420$, $\sigma = 8$ and $P(Z < z)$ = area shaded = 0.007.

Calculator gives answer for $w = 400.34$

(c) Looking at the standard normal distribution:

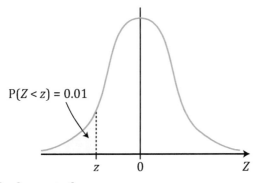

$P(Z < z) = 0.01$

z 0 Z

For the lower tail

Use the 'Inverse Normal' on the calculator with $\mu = 0$, $\sigma = 1$ and $P(Z < z)$ = area shaded = 0.01.

The calculator gives a z-value of -2.3263.

When $x = 400$ $z = \dfrac{x - \mu}{\sigma}$

Hence $-2.3263 = \dfrac{400 - \mu}{\sigma}$

$-2.3263\sigma = 400 - \mu$ (1)

> You can't use 'Inverse Normal' as you don't know the values for the mean or the standard deviation. Instead, you need to use the standardised normal distribution to find the z-values in each tail and then use the formula $z = \dfrac{x - \mu}{\sigma}$ in each case to find two equations in μ and σ that can then be solved simultaneously.

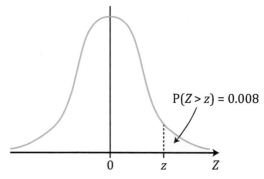

For the upper tail

Use the 'Inverse Normal' on the calculator with $\mu = 0$, $\sigma = 1$ and $P(Z > z) = 1 - \text{area shaded} = 1 - 0.008 = 0.992$

The calculator gives a z-value of 2.4089.

When $x = 440$ $z = \dfrac{x - \mu}{\sigma}$

Hence $2.4089 = \dfrac{440 - \mu}{\sigma}$

$$2.4089\sigma = 440 - \mu \qquad\qquad (2)$$

Solving equations (1) and (2) simultaneously gives:

$\sigma = 8.4474$ and $\mu = 419.65$

You can solve simultaneous equations using your Casio Classwiz calculator. It is worth learning how as it would save you a lot of time for this question.

Exam practice

1   Choose the most appropriate probability distribution to model these situations and give reasons for your choice.
   (a) The distribution of numbers obtained by the spinning of a fair hexagonal spinner numbered from 1 to 6, one hundred times. [2]
   (b) The distances travelled for a day trip in the summer. [2]
   (c) The wait in minutes at airport security. [2]
   (d) The salaries of 80 teachers at a school. [2]
   (e) The number of cuttings of a plant successfully growing when 200 cuttings are taken and the probability of a cutting successfully growing is 0.85. [2]
   (f) A continuous probability distribution which is bell-shaped and is symmetrical about the mean value. [2]

2 The time in minutes to check-in at an airport follows a continuous uniform distribution over the interval [3, 6].
  (a) A traveller arrives at the check-in. How long is their expected wait? [2]
  (b) Find the variance in minutes of their wait. [2]
  (c) Find the probability that a traveller waits between 3 and 5 minutes for check-in. [2]

3 A continuous random variable X is uniformly distributed over the interval [a, 10].
  (a) If $P(X \geq 8) = 0.2$, find the value of a. [2]
  (b) Write down the variance of X. [1]
  (c) Find $P(X \leq 5)$. [2]
  (d) Find $P(2 \leq X \leq 7)$. [2]

4 A string of length 10 cm is cut at a randomly chosen point. Find the probability that the length of the longer piece will be more than 7 cm. [3]

5 X is a random variable that can be modelled by a continuous uniform distribution over the interval [a, b], where $a > 0$ and $b > a$.
  (a) If the mean value of X is 5 and its variance is 3, find the values of a and b. [3]
  (b) Find $P(X \leq 5)$. [2]

6 X is a normally distributed variable with a mean of 200 and a variance of 16.
Find the value of x, correct to two decimal places such that:
  (a) $P(X < x) = 0.3$ [2]
  (b) $P(X > x) = 0.6$ [2]

7 The ages of customers of a restaurant are normally distributed with a mean of 34 years and a standard deviation of 11 years.
  (a) What is the probability that a randomly picked customer has an age of over 50 years? [2]
  (b) What is the probability that a randomly picked customer has an age under 21 years? [2]
  (c) What is the probability that a randomly picked customer is aged between 20 and 35 years? [3]

8 A machine fills packets with cereal. The machine is set to put a mean weight of 685 g of cereal with a standard deviation of 5 g into each packet.
  (a) Name a distribution that can be used to model the distribution of weights and give the parameters. [2]
  (b) If the cereal packets are marked with a weight of 675 g, find the probability that a randomly picked packet has a weight of less than 675 g. [3]

(c) The company packaging and selling the cereal could be prosecuted if the weight of cereal in each packet is less than 675 g. Find what the mean weight delivered by the machine should be adjusted to for there to be only a probability of 0.0001 of there being under 675 g. [5]

9. Customers ring up an IT helpline at a random point in time. Calls to the helpline take between 2 and 10 minutes for the customer to be given the help they need.

(a) Nicky works on the helpline and receives a call.
   (i) Suggest an appropriate distribution to model the time Nicky spends with each customer and give the parameters. [2]
   (ii) State the mean and variance of this distribution. [2]
   (iii) State an assumption you have made in suggesting this distribution. [1]

(b) Nicky is one of five people who answer the phones.
   (i) Find the probability that a caller to the helpline is answered by Nicky and the help given takes over 8 minutes. [3]
   (ii) Find the probability that a caller to the helpline is **not** answered by Nicky and the help given takes between 5 and 7 minutes. [2]

10. A farmer takes a random sample of 100 new potatoes from this year's harvest and records their masses in the following table:

Mass of potato (g)	$m < 25$	$25 \leq m < 35$	$35 \leq m < 45$	$45 \leq m < 55$	$55 \leq m < 65$	$65 \leq m < 75$	$m > 75$
Number of potatoes	0	3	29	36	30	2	0

(a) Explain why a normal distribution can be used to model the above data. [1]

The farmer models the above distribution of masses using $N(47.5, 10^2)$.

(b) (i) Using this model, find the probability that a potato picked at random from this sample has a mass of over 75 g and hence find the number of potatoes predicted having a mass over 75 g. [2]
   (ii) Using this model, find the number of potatoes in the sample that have a mass m g in the range $35 \leq m \leq 45$. [2]

(c) Explain by referring to your answers in part (b) whether the model suitably reflects the actual data in the table. [1]

(d) Could the model be used to predict the distribution of masses potatoes in future years? Explain your answer. [1]

⑪ The mean mark in a maths exam was 65 with a standard deviation of 6 marks.

(a) Explain why a normal distribution is appropriate for the modelling of the marks. [1]

(b) Find the probability that a student obtains over 75 marks. [2]

(c) 70 students took the exam. Find how many students passed the exam if the pass mark was 60 marks. Give your answer to the nearest integer. [3]

(d) The chief examiner has decided that a greater number of students should be passing so they decide to alter the pass mark so that 85% of students pass. Find this new pass mark giving your answer to the nearest integer. [3]

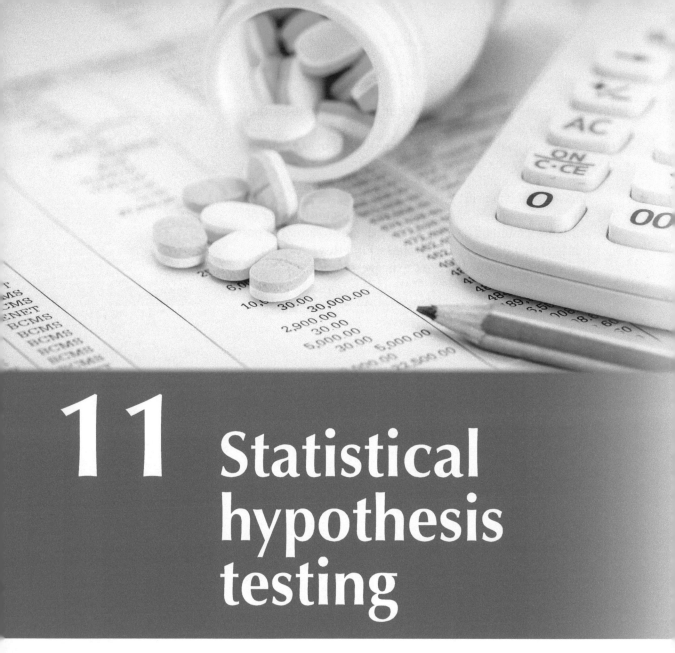

11 Statistical hypothesis testing

Prior knowledge

You will need to make sure you fully understand the following from your AS studies:

- Understanding and applying the language of hypothesis testing.

- Finding critical values and critical regions using hypothesis testing.

- One and two-tailed tests.

- Conducting a hypothesis test using *p*-values.

Quick revision

Quick revision

Correlation coefficients

Have the letters ρ if based on a population or r if based on a sample.

They have values between and including −1 and 1.

ρ or $r = 0$ means no correlation

ρ or $r = 1$ means perfect positive correlation

ρ or $r = -1$ means perfect negative correlation

Performing hypothesis testing using a correlation coefficient as a test statistic

Testing for positive correlation use $\mathbf{H_0} : \rho = 0, \ \mathbf{H_1} : \rho > 0$

Testing for negative correlation use $\mathbf{H_0} : \rho = 0, \ \mathbf{H_1} : \rho < 0$

Testing for any correlation (i.e. positive or negative) use
$\mathbf{H_0} : \rho = 0, \ \mathbf{H_1} : \rho \neq 0$

Hypothesis testing for the mean of a normal distribution with a known, given or assumed variance

If X is normally distributed then $X \sim N(\mu, \sigma^2)$ and the sample mean, $\overline{X}$, is normally distributed, so $\overline{X} \sim N\left(\mu, \frac{\sigma^2}{n}\right)$

If $Z \sim N(0, 1)$, then $\qquad Z = \dfrac{\overline{X} - \mu}{\frac{\sigma}{\sqrt{n}}}$

> This formula is used to work out z-values that apply to the standard normal distribution.

Hence the z-value (i.e. the test statistic) will be:

$$z = \dfrac{\overline{X} - \mu}{\frac{\sigma}{\sqrt{n}}}$$

There are two methods that can be used for hypothesis testing:

- Using the z-value to find the probability (i.e. the p-value) and then compare this probability to the significance level.

- Use the significance level of the test to find the critical value and then see if the test statistic (i.e. the sample mean) lies inside or outside the critical region.

Performing a hypothesis test using *p*-values

A *p*-value is simply a probability. For example, $P(X > 19) = 0.08$ means the *p*-value is 0.08.

Once a *p*-value has been found, it is compared against the significance level (usually 1% or 5%).

If the **p-value < significance level**, there is **evidence to reject the null hypothesis**.

If the **p-value > significance level**, there is **evidence to fail to reject the null hypothesis**.

For a two-tailed test, you must remember that you must add the *p*-value for each tail (which is usually the same) together to give the final *p*-value. You then compare this with the significance level.

Remember: Don't say 'accept the null hypothesis' say instead 'fail to reject the null hypothesis'.

Also, always state your result in context. Explain what the result means to the person conducting the test.

Performing a hypothesis test using critical values

To perform a hypothesis test using critical values:

Using the significance level, you can work out the critical value using a calculator.

To do this use Inverse Normal and enter the following parameters:

- The area which will be the significance level as a decimal (i.e. 5% = 0.05, 1% = 0.001)

  If you are performing a two-tailed test, you need to halve the significance level with each area at the end of each tail.

  It is also important to note that when you are finding a value of *x* using Inverse Normal the area is measured from the lower tail towards the upper tail.

- The standard deviation (remember that if a sample is being used this will be the standard deviation of the population divided by the square root of the sample size (i.e. $\frac{\sigma}{\sqrt{n}}$))

- The mean (note that this will be the original mean).

The answer gives you the critical value which can then be compared with the sample mean. If the sample mean is in the critical region the result is significant so there is evidence to reject the null hypothesis.

Looking at exam questions

1   The heights of a type of plant are normally distributed with a mean height of 18 cm and a standard deviation of 3 cm. With the addition of a new fertiliser, the plants are claimed to grow taller. 50 plants were grown using the new fertiliser and had a mean height of 19 cm. Test at the 5% significance level what conclusion can be drawn about the addition of the fertiliser.   [10]

Thinking about the question

This question is concerned with hypothesis testing. We can either use an approach by making use of p-values, or we can find the critical value. In this case we will find the critical value and then the critical region and see if the new height after using fertiliser lies inside or outside the critical region.

Starting the solution

We can start off by stating the null and alternative hypotheses. The null hypothesis is the status quo (i.e. nothing has changed so the mean height remains at 18 cm) and the alternative hypothesis is that the height is greater than 18 cm. Because we are only investigating one tier of the normal distribution, this is a one-tailed test.

We can then use Inverse Normal on the calculator to work out the value of $\bar{x}$, which will be the critical value, for a probability of 0.95.

We can then do a sketch of the distribution showing the critical value and shading the critical region. We can also mark on the value of $\bar{x}$ to see if it lies in the critical region. If it lies in the critical region then there is sufficient evidence to reject the null hypothesis. We need to ensure that we write our conclusion in context.

The solution

$$H_0 : \mu = 18 \qquad H_1 : \mu > 18$$

Let X be the mean height of plants which is normally distributed.

Assuming H_0, if $X \sim N(\mu, \sigma^2)$ then the sample mean $\bar{X} \sim N\left(\mu, \frac{\sigma^2}{n}\right)$

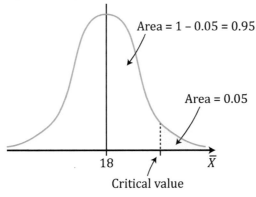

Area = 1 – 0.05 = 0.95

Area = 0.05

18

$\bar{X}$

Critical value

> Remember that when finding the area to use with Inverse Normal we must measure the area from the lower tail. Hence Area = 1 – 0.05 = 0.95.

Use Inverse Normal with the following parameters:

Area: 0.95 (i.e. 1 – 0.05 = 0.95)

$\sigma: \dfrac{3}{\sqrt{50}} = 0.4242$

μ: 18

Gives a critical value of 18.6979.

The critical region is $\overline{X} \geq 18.6979$

The sample mean = 19, so it lies in the critical region, so the result is significant.

> The null hypothesis is rejected in favour of the alternative hypothesis.

Hence there is sufficient evidence at the 5% level of significance to conclude that adding the fertiliser increases the height of the plants.

2   A company produces kettlebells whose weights are normally distributed with mean 16 kg and standard deviation 0.08 kg.

(a) Find the probability that the weight of a randomly selected kettlebell is greater than 16.05 kg.   [2]

(b) The company trials a new production method. It needs to check that the mean is still 16 kg. It assumes that the standard deviation is unchanged. The company takes a random sample of 25 kettlebells and it decides to reject the new production method if the sample mean does not round to 16 kg to the nearest 100 g.

Find the probability that the new production method will be rejected if, in fact, the mean is still 16 kg.   [4]

(c) The company decides instead to use a 5% significance test. A random sample of 25 kettlebells is selected and the mean is found to be 16.02 kg.

Carry out the test to determine whether or not the new production method will be rejected.   [6]

(WJEC June 2019 Unit 4 q4)

Thinking about the question

This question is about working out probabilities using the normal distribution. The easiest way to work these out is to use a calculator. For part (a) we have to find $P(X > 16.05)$. For part (b) we find the range of values for which the new production would be rejected and then find the probability of this happening. For part (c) we need to perform a hypothesis test.

Starting the solution

For part (a) we can use the Normal CD on the calculator. It is worth drawing a quick sketch to show the area which represents the probability being found.

For part (b) we need to first think about the limits for which the new production method would be rejected. 16 kg to the nearest 0.1 kg (i.e. 100 g) would place these limits at 16.05 and 15.95. This means the sample mean would be either less than 15.95 or greater than 16.05 for the new production method to be rejected. As the curve is symmetrical This means we only need to find one area and then we can simply double it, as the other area will be the same. We need to remember that we are looking at a normal distribution of a sample mean, $\overline{X}$ so the mean is 16 and the variance is now $\frac{\sigma^2}{n}$, so $\overline{X} \sim N\left(16, \frac{0.08^2}{25}\right)$. The standard deviation is therefore $\dfrac{0.08}{5} = 0.016$. We can then use the normal distribution along with the mean and standard deviation to work out the areas.

For part (c) we need to perform a hypothesis test. We again need to be aware that we are using a sample and not a population. The null hypothesis is that that the mean is 16 and the alternative hypothesis is that the mean is not 16, so this is a two-tailed test.

The solution

(a) Let the random variable X be the weights of the kettlebells.

X is normally distributed with mean 16 kg and standard deviation 0.08 kg.

So, $X \sim N(16, 0.08^2)$

> Note that the question implies that the population is being used in part (a).

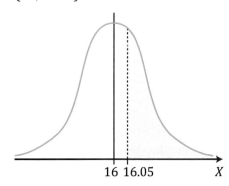

To find $P(X > 16.05)$ using the CASIO Classwiz calculator, take the following steps:

Select 'Distribution' and then 'Normal CD'

Now enter the following parameters:

Lower: 16.05

Upper: 1×10^{99}

σ: 0.08

μ: 16

The answer is displayed as P = 0.26598....

Probability, p-value = 0.26599

(b) $\bar{X} \sim N\left(16, \frac{0.08^2}{25}\right)$

> We must remember here that we are hypothesis testing the mean of a normal distribution using a sample.

16 kg to the nearest 0.1 kg means the values could be <15.95 kg or >16.05 kg for the new production method to be rejected.

Notice the areas required in upper and lower tails are the same.

$P(\bar{X} < 15.95)$ $P(\bar{X} > 16.05)$

15.95   16   16.05 $\bar{X}$

To find $P(\bar{X} < 15.95)$ we use Normal CD with the parameters:

Lower: 1×10^{-99}

Upper: 15.95

σ: $\dfrac{0.08}{5} = 0.016$

μ: 16

The answer is displayed as P = 8.8903×10^{-4}

Probability = $P(\bar{X} < 15.95) + P(\bar{X} > 16.05)$

$\qquad\qquad = 2 \times 8.8903 \times 10^{-4}$

$\qquad\qquad = 0.00178$

Probability production method is rejected = 0.00178

> To find $P(\bar{X} > 16.05)$ we can use the fact that the two areas being found are the same, as the normal distribution curve is symmetrical. So, you can simply double the area found for $P(\bar{X} < 15.95)$.

(c) This part of the question will be solved using *p*-values.

$$H_0 : \mu = 16 \qquad H_0 : \mu \neq 16$$

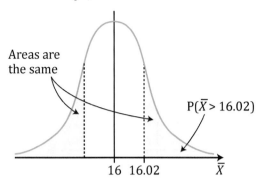

Areas are the same

$P(\overline{X} > 16.02)$

16   16.02 $\overline{X}$

> Note that this is a two-tailed test.

For $P(\overline{X} > 16.02 \backslash H_0)$

we use Normal CD with the parameters:

Lower: 16.02

Upper: 1×10^{99}

$\sigma: \dfrac{0.08}{5} = 0.016$

$\mu: 16$

The answer is displayed as P = 0.10565

p-value = 2 × 0.10565

= 0.2113

> This means the probability the sample mean of X is greater than 16.02 given the null hypothesis is true (i.e. $\mu = 16$).

> As this is a two-tailed test the probability in one of the tails must be doubled to give the total probability (i.e. the *p*-value).

Now 0.2113 > 0.05

As the *p*-value is greater than the significance level there is insufficient evidence to reject the null hypothesis H_0.

Hence, there is sufficient evidence that the new production method should be adopted.

> The *p*-value is the test statistic, and this is compared against the significance level (5% = 0.05). If the *p*-value is greater than the significance level, there would be insufficient evidence to reject the null hypothesis.

This is the same part of the question answered using critical values

(c)

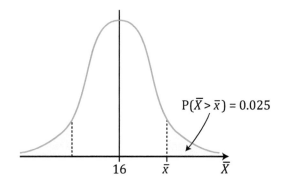

$P(\bar{X} > \bar{x}) = 0.025$

16    $\bar{x}$ $\bar{X}$

To find the area we must remember that it is measured starting from the lower tier towards the upper tier. Hence the area we need to use is 1 – 0.025 = 0.975

Remember that the area when you use Inverse Normal is measured from the lower tail. An area of 0.01 measured in the upper tail is 0.99 measured from the lower tail.

Using Inverse Normal with the following parameters:

Area: 0.975

$\sigma: \dfrac{0.08}{\sqrt{25}} = 0.016$

μ: 16

Gives the critical value $\bar{x} = 16.031$

Now the sample mean = 16.02

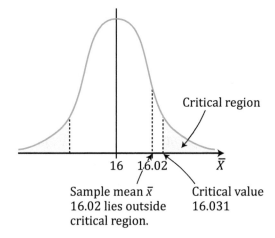

16   16.02 $\bar{X}$

Sample mean $\bar{x}$ Critical value
16.02 lies outside   16.031
critical region.

Critical region

As the sample mean < critical value it lies outside the critical region, so there is insufficient evidence to reject the null hypothesis H_0.

Hence, there is sufficient evidence that the new production method should be adopted.

3 A bowling alley manager in the UK is concerned about falling revenues. He collects data from the United States, hoping to use what he finds to revive his business in the UK.

He finds data which seem to show correlation between margarine consumption and bowling alley revenue. He attempts to carry out some statistical analysis in order to present his findings to the board of directors. He produces the scatter diagram shown below:

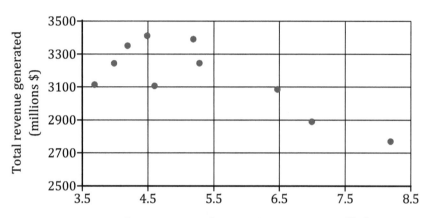

Consumption of margarine versus total revenue generated by bowling alleys

The product moment correlation coefficient for these data is –0.7617. He carries out a one-tailed test at the 1% level of significance and concludes that higher margarine consumption is associated with lower revenue generated by bowling alleys.

(a) Show all the working for this test. [5]

The manager also conducts a significance test for bowling alley revenue and fish consumption per person. He produces the computer output, shown below, for the analysis of bowling alley revenue versus fish consumption per person.

```
# Pearson's product-moment correlation
# data: revenue and fish
# t = 3.8303, df = 8, p-value = 0.005215
# alternative hypothesis: true correlation is
not equal to 0
# sample estimates:
# correlation
# 0.8024423
```

(b) Comment on the correlation between bowling alley revenue and fish consumption per person and what the board of directors should do in light of the manager's findings in part (a) and part (b). [3]

(c) Give one possible reason why the board of directors might not be happy with the manager's analysis. [1]

(WJEC June 2019 Unit 4 q5)

Thinking about the question

On reading this question the first thing to remember is that correlation does not imply causation. It is a bit bizarre that anyone would think the quantities mentioned in the question are causal. Part (a) involves hypothesis testing and part (b) involves looking at the manager's analysis.

Starting the solution

For part (a) we need to write the null and alternative hypothesis and then use $n = 10$ (i.e. the number of points on the graph) with the test statistic −0.7617 to determine the critical value. The critical value is then compared with the test statistic to see if the null hypothesis should be rejected.

For part (b) we use the p-value from the computer output and compare it with the significance level. As the significance level is 1% (i.e. 0.01) if the p-value is less than this, then it will cause the null hypothesis (that there is no correlation) to be rejected.

For part (c) we need to study the graph and try to understand what each quantity on the axes represents and then what the points mean.

We notice that no information is given to say whether the data is the population or a sample.

The solution

(a) Null hypothesis $H_0 : \rho = 0$ (i.e. no correlation)

Alternative hypothesis $H_1 : \rho < 0$ (i.e. negative correlation)

The test statistic is the product moment correlation coefficient which is −0.7617

We now find the critical value of the product moment correlation coefficient (PMMC) using Table 9.

We are performing a one-tailed test using 10 pairs of observations. So, we look up a 1% one-tailed test with a value of $n = 10$ and this gives a critical value of 0.7155. Critical values can be positive or negative, so it is easier to simply change the PMMC to +0.7617.

Now, the test statistic (PMCC 0.7167) > the critical value (0.7155) so there is sufficient evidence to reject H_0. Hence, there is evidence that his conclusion is correct.

(b) The p-value of 0.005215 from the results is less than the significance level of 0.01 (i.e. 1%). This means that there is evidence to reject the null hypothesis.

The correlation of 0.8024423 means the bowling alley revenue is positively correlated with fish consumption per person.

(c) Although the two quantities are significantly correlated, there is no sense in the result so the board of directors should not implement any of the suggestions made by the manager based on his calculations.

There are some alternative answers to the answer above and they include:

We do not know whether the data is a sample, so it is hard to draw conclusions.

He has carried out the test without considering how irrelevant the results will be.

No mention of what the population is has been made.

Exam practice

1. A firm produces lengths of cord with a mean breaking strength of 20 Newtons and standard deviation 0.769. It is claimed that if the cord is treated with a new chemical, the mean breaking strength increases. To test this claim, a sample of 10 lengths of cord were selected at random and treated with the new chemical. The mean breaking strength of the sample was found to be 20.6.
   (a) State suitable hypotheses for testing this claim.   [1]
   (b) Carry out an appropriate test with significance level 1% and state your conclusion in context. Explain how you reached your conclusion.   [5]

(WJEC June 18 S3 q3 changed question)

2   A farmer suspects that older chickens produce fewer eggs, so
    she collects some data about ages of chickens and the number
    of eggs produced for the chickens on her farm.
    The farmer uses the number of eggs and age of 15 chickens
    to calculate the product moment correlation coefficient and
    finds it to be −0.87. The farmer has decided to test for negative
    correlation.
    (a) State the null and alternative hypotheses the farmer
         could use for this test.   [1]
    (b) Using the 'Critical values of the product moment
         correlation coefficient table' carry out a hypothesis test at
         the 5% level of significance and state whether the results
         are significant and write a conclusion in context.   [6]

3   The mean weight of a certain breed of bird is believed to be
    4.5 kg. In order to test this belief, a random sample of 10 birds
    of the breed was obtained and weighed, and the mean weight
    was found to be 4.2 kg.
    You may assume that the weights of this breed of bird are
    normally distributed with a standard deviation of 0.1656.
    (a) State suitable hypotheses for testing the above belief using
         a two-sided test.   [1]
    (b) Carry out an appropriate test using a 1% significance
         level and state your conclusion in context, justifying your
         answer.   [7]

4   A company sells cans of beer and the volume of beer in each
    can has a normal distribution with a standard deviation of
    8 ml. The company say that the mean volume of beer in the
    can is 445 ml, but some customers have complained to trading
    standards that the volume is less than this. The officer from the
    trading standards department takes a random sample of 40
    cans and finds that the mean volume of beer in each can is 443 ml.
    (a) Explain why a normal distribution can be used to model
         the volume of liquid in the cans.   [1]
    (b) State two hypotheses that could be used to determine if the
         volume of beer in each can is less than that stated by the
         company.   [1]
    (c) Test at the 5% level of significance whether there is
         evidence the customer complaints are justified.   [4]

⑤ Suraiya collects data concerning the average number of hours of TV watched per day and the average IQ for 10 different countries to investigate whether there is any correlation. Suraiya uses the data to plot the following scatter graph:

**How IQ changes with mean number
of hours of TV watched per day**

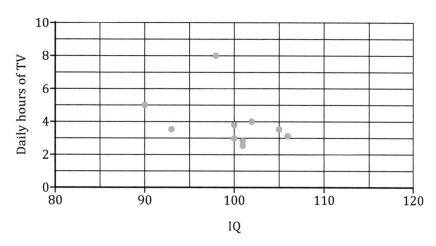

The product moment correlation coefficient for these data is –0.36. Suraiya carries out a one-tailed test at the 1% level of significance and concludes that higher intelligence is **not** associated with lower number of daily hours of TV viewing.

(a) Comment on the correlation between daily hrs of TV watched and IQ. [2]

(b) Show all the working for this test. [5]

⑥ The time for treatment by a dentist can be modelled by a normal distribution with a standard deviation of 5 minutes. Given that the probability of the treatment taking more than 40 minutes is 0.04, find:

(a) The mean treatment time. [3]

(b) The probability of taking more than 30 minutes. [2]

(c) The manufacturer of a new piece of dental equipment has claimed that it saves time. The dentist trials this equipment with a sample of 20 patients and the mean treatment time using this sample is 30 minutes and the standard deviation is still 5 minutes.
Test at the 5% significance level, stating clearly the null and alternative hypotheses, to determine whether there is sufficient evidence that the manufacturer's claim is true. [5]

7 A machine makes metal coasters with a diameter D where D is distributed normally with a mean of 9 cm and a standard deviation of 0.35 cm.

The production manager suspects the machine is faulty and is not making coasters of the correct diameter. She takes a random sample of 20 coasters to see if the diameter has changed from 9 cm.

(a) (i) Write down two suitable hypotheses to test if the diameters have changed. [1]

   (ii) Find at the 1% level of significance, the critical region for the test. [3]

(b) The mean diameter of the sample of 20 coasters was found to be 9.1 cm.

Explain the significance of this result to the production manager. [2]

12 Kinematics for motion with variable acceleration

Prior knowledge

You will need to make sure you understand the following from your AS studies:

- The terms used in kinematics (position, displacement, distance travelled, velocity, speed and acceleration).

- Displacement–time graphs
- Velocity–time graphs.
- Sketching and interpretation of velocity–time graphs.

Quick revision

Using calculus when the acceleration is not constant

Calculus has to be used to find s, v, or a when the acceleration is not constant, and the following diagrams shows the processes involved.

Using differentiation

Displacement (s) $\xrightarrow[\dfrac{ds}{dt}]{\text{Differentiate}}$ Velocity (v) $\xrightarrow[\dfrac{dv}{dt}]{\text{Differentiate}}$ Acceleration (a)

Using integration

Displacement (s) $\xleftarrow[s = \int v\,dt]{\text{Integrate}}$ Velocity (v) $\xleftarrow[v = \int a\,dt]{\text{Integrate}}$ Acceleration (a)

Formulae

$$v = \frac{ds}{dt}$$

$$a = \frac{dv}{dt}$$

$$s = \int v\,dt$$

$$v = \int a\,dt$$

Important facts

The gradient of a displacement–time graph is the velocity.

The gradient of a velocity–time graph is the acceleration.

The area under a velocity–time graph is the displacement.

Newton's 2nd law of motion

A resultant force, F N, produces acceleration, a ms^{-2}, on a mass, m kg, according to the formula: Force = mass × acceleration or $F = ma$.

Looking at exam questions

1 A particle moves in a straight line and its velocity v ms^{-1}, t seconds after passing a point O, is given by the equation:

$$v = 6 + 3t^2.$$

Find the distance travelled between the times $t = 1$s and $t = 2$s. [3]

Thinking about the question

Looking at the equation you can see that the velocity depends on time. This means the velocity is not constant. Hence, we need to use calculus rather than the equations of motion.

Starting the solution

The distance travelled between the two times is equal to the area under the velocity time graph between these two times. We therefore integrate the velocity equation with respect to t using the two times (i.e. 2 and 1) as the limits.

The solution

$$s = \int_1^2 (6 + 3t^2)\, dt$$

$$= [6t + t^3]_1^2$$

$$= [(12 + 8) - (6 + 1)]$$

$$= 13\ m$$

> Use the formula $s = \int v\, dt$ which needs to be remembered.

---

2 Particle P having mass 0.5 kg, moves along the x-axis so that its velocity, v ms^{-1} at time t s is given by:

$$v = 2t^3 - 2t + 1.$$

When $t = 0$s, the displacement of the particle is 4 m from the origin.

(a) Find the displacement of P from the origin when $t = 2$s. [4]

(b) Find the force that acts on the particle when $t = 2$s. [3]

Thinking about the question

The velocity varies with time, so we need to use calculus. For part (a) we can integrate the velocity and for part (b) we need to find the acceleration first by differentiating the velocity equation.

Starting the solution

For (a) we integrate the velocity to find the expression for the displacement which will involve a constant of integration. To find the constant we can use the information that when $t = 0$ s, $s = 4$ m. We then substitute the value for c back into the equation for s and then substitute $t = 2$ s to obtain the value for the displacement.

For part (b) we differentiate v to find an equation for a. We then substitute $t = 2$ into this equation to find a value for a. The formula $F = ma$ is then used to find the force, F.

The solution

(a) $s = \int v\,dt$

$$= \int (2t^3 - 2t + 1)\,dt$$

$$= \frac{2t^4}{4} - \frac{2t^2}{2} + t + c$$

$$= \frac{t^4}{2} - t^2 + t + c$$

When $t = 0$, $s = 4$.

$$4 = \frac{(0)^4}{2} - (0)^2 + (0) + c$$

$$c = 4$$

Hence, $s = \frac{t^4}{2} - t^2 + t + 4$

When $t = 2$, $s = \frac{(2)^4}{2} - (2)^2 + (2) + 4 = 10$

Displacement $= 10\,m$

(b) $v = 2t^3 - 2t + 1$.

To find the force we need to find the acceleration first by differentiating v. We can then use $F = ma$ to find the force.

$$a = \frac{dv}{dt}$$

$$= 6t^2 - 2$$

When $t = 2s$, $a = 6(2)^2 - 2 = 22\,ms^{-2}$

$F = ma = 0.5 \times 22 = 11\,N$

---

3 A particle of mass 4 kg moves along the x-axis, starting, when $t = 0$, from the point where $x = 3$. At time t s, its velocity $v\,ms^{-1}$ is given by:

$$v = 12t^2 - 7kt + 1, \text{ where } k \text{ is constant.}$$

When $t = 2$, the displacement of the particle from the origin is 16 m.

(a) Determine the value of k. [5]

(b) Calculate the magnitude of the force acting on the particle when $t = 5$. [4]

(WJEC June 2016 M2 q1)

Thinking about the question

We can see that the question contains information about the displacement at a certain time. If we integrate the velocity, we obtain the displacement. On integration we will now have two constants, the constant of integration, c and the constant k. We will have to find c and hence find k.

To find the force on the particle we need to find the acceleration. An equation for the acceleration can be found by differentiating the velocity equation.

Starting the solution

For part (a), after integrating we can find the value of the constant of integration by substituting $t = 0$ and $s = 3$ into the equation and solving. We then substitute this value back and then substitute $t = 2$ and $s = 16$ and solve the resulting equation to find k.

For part (b) we can differentiate v to obtain the acceleration, a. By substituting $t = 5$ into this equation we can find the value of the acceleration a. We can then recall and use the formula $F = ma$ to find the value for the force on the particle at this time.

The solution

(a) $s = \int v \, dt$

$\quad = \int (12t^2 - 7kt + 1) dt$

$\quad = 4t^3 - \dfrac{7kt^2}{2} + t + c$

When $t = 0, s = 3$

$\quad 3 = 0 - 0 + 0 + c$ so $c = 3$

Hence, we have $s = 4t^3 - \dfrac{7kt^2}{2} + t + 3$

When $t = 2, s = 16$

$\quad 16 = 4(2)^3 - \dfrac{7k(2)^2}{2} + 2 + 3$

$\quad 16 = 32 - 14k + 2 + 3$

$\quad 14k = 21$

$\quad k = \dfrac{3}{2}$

(b) $v = 12t^2 - \dfrac{21}{2}t + 1$

$\quad a = \dfrac{dv}{dt}$

$\quad = 24t - \dfrac{21}{2}$

> Notice that there are now two constants, the constant in the original equation, k and the constant of integration, c.

> We now use the values given to find the values of the constants, c and k.

$$\text{When } t = 5\text{s}, a = 24(5) - \frac{21}{2} = 109.5$$

$$F = ma = 4 \times 109.5 = 438 \text{ N}$$

4 A particle moves in a straight line with velocity v ms^{-1} at time t s given by:

$$v = 3 \cos 2t$$

Calculate the distance travelled by the particle between $t = 0$ and $t = \frac{\pi}{2}$ s.   [5]

Thinking about the question

To find displacement we integrate the velocity equation. We note that this question asks for the distance rather than the displacement.

Starting the solution

The distance can be found by finding the area between the curve and the time axis. We need to be careful here as some of the area under the x-axis is negative. We require the distance, so we are only interested in the area and not in its sign. We need to draw a graph of the velocity to see which parts of the graph will result in a negative area.

The solution

$$v = 3 \cos 2t$$

As the displacement is the area under the velocity–time graph, we need to draw a graph to show how velocity varies with time. We can see that if we integrated between $t = 0$ and $t = \frac{\pi}{2}$, this would give a displacement of zero. As the area above the x-axis which is positive would be added to area below the x-axis which is negative. As we want the distance and not the displacement, we can find the area between $t = 0$ and $t = \frac{\pi}{4}$ and then double it owing to the symmetry of the graph.

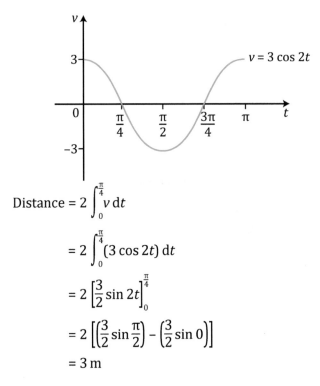

$$\text{Distance} = 2 \int_0^{\frac{\pi}{4}} v \, dt$$

$$= 2 \int_0^{\frac{\pi}{4}} (3 \cos 2t) \, dt$$

$$= 2 \left[\frac{3}{2} \sin 2t \right]_0^{\frac{\pi}{4}}$$

$$= 2 \left[\left(\frac{3}{2} \sin \frac{\pi}{2} \right) - \left(\frac{3}{2} \sin 0 \right) \right]$$

$$= 3 \text{ m}$$

Exam practice

1. A particle moves along the positive x-axis such that its velocity, v ms^{-1}, at time t s, is given by:
$$v = 9t^2 - 6t + 1$$
The particle is at the origin, O, at $t = 0$ s.
Find the distance from O when the particle is moving with its minimum velocity. [6]

2. Particle P moves along the x-axis such that at time t s, the velocity v ms^{-1} of P is given by:
$$v = 2t(2 - t)(1 - t)$$
   (a) Find an expression for the acceleration of the particle at time t s. [2]
   (b) Given that when $t = 0$, $s = 3$, find the displacement of the particle at time $t = 2$ s. [4]

3. A particle P moves along the x-axis and at t s it has a velocity v ms^{-1} in the direction of the positive x-axis where:
$$v = 2t^2 - 18t + 36$$
   (a) Find the times when the particle is at instantaneous rest. [2]
   (b) Find the time when the particle velocity is least and find this minimum velocity. [3]
   (c) Find the total distance travelled by the particle between $t = 2s$ and $t = 6s$. [5]

4. A particle moves in a straight line such that its acceleration a ms^{-2} is given by:
$$a = 4 - 6t \qquad \text{for } t \geq 0.$$
At time $t = 0$, the particle is at the point O and its velocity is 4 ms^{-1}.
   (a) Find an expression for the velocity of the particle at time t s. [3]
   (b) Find an expression for the displacement of the particle from O at time t s. [3]
   (c) Determine the time when the particle comes to rest instantaneously and the distance of the particle from O at this time. [3]
   (d) Calculate the speed of the particle when $t = 3$, and determine whether or not the speed of the particle is increasing or decreasing at this time. [3]

(WJEC June 2005 M2 q2)

5 A particle of mass 2 kg, moves along the horizontal x-axis under the action of a constant force, F N. The velocity, v ms^{-1}, after t s is given by:

$$v = 8t - 18t^2$$

(a) The particle is at the origin when $t = 1$s. Find an expression for the displacement of the particle from the origin. [4]

(b) Find the acceleration of the particle at time t s. [2]

6 A particle moves in a straight line with velocity, v ms^{-1} at time t s, where:

$$v = -3 \sin \frac{t}{2}$$

(a) Draw a velocity–time graph of v for $0 \le t \le \pi$. [3]

(b) Calculate the distance travelled by the particle between $t = 0$ and $t = \dfrac{\pi}{2}$ s. [3]

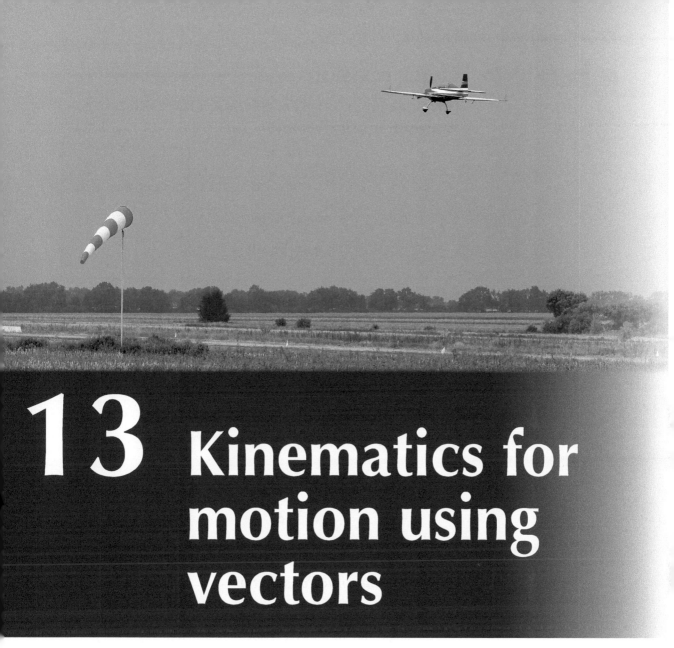

13 Kinematics for motion using vectors

Prior knowledge

You will need to make sure you fully understand the following from your AS studies:

- Calculating the magnitude and direction of a vector and converting between component form and magnitude/direction form.

- Using vectors to solve problems.

Quick revision

Formulae for constant acceleration for motion in a straight line using vectors

$$\mathbf{v} = \mathbf{u} + \mathbf{a}t$$

$$\mathbf{s} = \mathbf{u}t + \frac{1}{2}\mathbf{a}t^2$$

$$\mathbf{v}^2 = \mathbf{u}^2 + 2\mathbf{a}\mathbf{s}$$

$$\mathbf{s} = \frac{1}{2}\left(\mathbf{u} + \mathbf{v}\right)t$$

> $\mathbf{s}$ = displacement
> $\mathbf{u}$ = initial velocity
> $\mathbf{v}$ = final velocity
> $\mathbf{a}$ = acceleration
> t = time

Newton's 2nd law of motion

Force = mass × acceleration or $\mathbf{F} = m\mathbf{a}$

Using calculus for motion in a straight line using vectors

For questions involving variable acceleration we make use of the following equations:

$$\mathbf{v} = \frac{d\mathbf{s}}{dt}$$

$$\mathbf{a} = \frac{d\mathbf{v}}{dt}$$

$$\mathbf{s} = \int \mathbf{v}\, dt$$

$$\mathbf{v} = \int \mathbf{a}\, dt$$

Finding the magnitude of a vector

- The magnitude of displacement is distance (a scalar).
- The magnitude of velocity is speed (a scalar).
- The magnitude of acceleration or force does not have its own name.
- For a vector in two dimensions such as vector, $\mathbf{v} = a\mathbf{i} + b\mathbf{j}$:

  magnitude of vector, $|\mathbf{v}| = \sqrt{a^2 + b^2}$
- For a vector in three dimensions such as vector, $\mathbf{v} = a\mathbf{i} + b\mathbf{j} + c\mathbf{k}$:

  magnitude of vector, $|\mathbf{v}| = \sqrt{a^2 + b^2 + c^2}$

Looking at exam questions

1   At time t, the position vectors relative to a fixed origin O, of two particles A and B are given by OA = $2\mathbf{i} + 3\mathbf{j} + \mathbf{k} + t(2\mathbf{i} - 6\mathbf{j} + 9\mathbf{k})$ and OB = $5\mathbf{i} - 8\mathbf{j} + 10\mathbf{k} + t(3\mathbf{i} - 6\mathbf{j} + 7\mathbf{k})$.

(a)   Find the speed of particle A.   [3]

(b)   Show that the distance AB at time t is given by:

$AB^2 = 5t^2 - 30t + 211$.

Determine the time at which the particles A and B are closest together.   [7]

(WJEC June 2011 M2 q7)

Thinking about the question

The two position vectors for particles at points A and B are given in the question. Notice that each position vector is made up of two parts. One part depends on time t and the other part doesn't. The position vector after time t s is the sum of the position vector and the displacement vector.

Starting the solution

For part (a) we need to identify the displacement vector for OA and as displacement = velocity × time, the part of the position vector being multiplied by the time will equal the velocity vector. The speed will be the magnitude of the velocity vector (i.e. $|\mathbf{v}|$).

For part (b) we need to first work out the vector **AB** and then square it to give AB^2.

We now need to find the smallest value of AB^2. We can find this by differentiating and then equating the differential to zero and then solving the resulting equation for t.

The solution

(a) $\mathbf{v} - 2\mathbf{i} - 6\mathbf{j} + 9\mathbf{k}$

$|\mathbf{v}| = \sqrt{2^2 + (-6)^2 + 9^2} = 11$

(b) $\mathbf{AB} = \mathbf{OB} - \mathbf{OA}$

$= 5\mathbf{i} - 8\mathbf{j} + 10\mathbf{k} + t(3\mathbf{i} - 6\mathbf{j} + 7\mathbf{k}) - [2\mathbf{i} + 3\mathbf{j} + \mathbf{k} + t(2\mathbf{i} - 6\mathbf{j} + 9\mathbf{k})]$

$= 3\mathbf{i} + t\mathbf{i} - 11\mathbf{j} + 9\mathbf{k} - 2t\mathbf{k}$

$= (3 + t)\,\mathbf{i} - 11\mathbf{j} + (9 - 2t)\,\mathbf{k}$

$AB^2 = (3 + t)^2 + (-11)^2 + (9 - 2t)^2$

$= 9 + 6t + t^2 + 121 + 81 - 36t + 4t^2$

$= 5t^2 - 30t + 211$

$\dfrac{dAB^2}{dt} = 10t - 30$

> Here we need to find the minimum value of $\frac{dAB^2}{dt}$. To find this, we equate $\frac{dAB^2}{dt}$ to zero.

At the minimum distance for AB, $\frac{dAB^2}{dt} = 0$

Hence, $10t - 30 = 0$

$$10t = 30$$

$$t = 3\,\text{s}$$

---

2   A particle P of mass 0.5 kg moves on a horizontal plane such that its velocity vector $\mathbf{v}$ ms^{-1} at time t seconds is given by:

$$\mathbf{v} = 12\cos(3t)\,\mathbf{i} - 5\sin(2t)\,\mathbf{j}.$$

(a)   Find an expression for the force acting on P at time t s.   [3]

(b)   Given that when $t = 0$, P has position vector $(4\mathbf{i} + 7\mathbf{j})$ m relative to the origin O, find an expression for the position vector of P at time t s.   [4]

(c)   Hence determine the distance of P from O at time $t = \dfrac{\pi}{2}$.   [2]

(WJEC June 2019 Unit 4 q6)

Thinking about the question

For part (a) we need to find the force, so we first need to determine the acceleration and then use the formula $\mathbf{F} = m\mathbf{a}$. For part (b) we need to find the displacement vector. For part (c) we substitute $t = \dfrac{\pi}{2}$ into the position vector equation obtained in part (b).

Starting the solution

For part (a) we differentiate the velocity in order to obtain the acceleration vector and then substitute $\mathbf{a}$ into $\mathbf{F} = m\mathbf{a}$ in order to obtain $\mathbf{F}$.

For part (b) we integrate the velocity to find the displacement and we can substitute the position vector and $t = 0$ into this in order to find the constant of integration. Once found we can write the position vector at t s.

For part (c) we substitute $t = \dfrac{\pi}{2}$ into the expression for the position vector obtained in part (b). The distance can then be obtained from the position vector.

The solution

(a) $\mathbf{a} = \dfrac{d\mathbf{v}}{dt} = -36\sin 3t\,\mathbf{i} - 10\cos 2t\,\mathbf{j}$

$\mathbf{F} = m\mathbf{a}$

$= 0.5(-36\sin 3t\,\mathbf{i} - 10\cos 2t\,\mathbf{j})$

$= -18\sin 3t\,\mathbf{i} - 5\cos 2t\,\mathbf{j}$

(b) $\mathbf{s} = \int \mathbf{v}\, dt$

$= \int (12\cos(3t)\,\mathbf{i} - 5\sin(2t)\,\mathbf{j})\, dt$

$= \dfrac{12\sin 3t}{3}\,\mathbf{i} + \dfrac{5\cos 2t}{2}\,\mathbf{j} + c$

$= 4\sin 3t\,\mathbf{i} + 2.5\cos 2t\,\mathbf{j} + c$

When $t = 0$, $\mathbf{s} = 4\mathbf{i} + 7\mathbf{j}$

$4\mathbf{i} + 7\mathbf{j} = 4\sin(0)\,\mathbf{i} + 2.5\cos(0)\,\mathbf{j} + c$

$4\mathbf{i} + 7\mathbf{j} = 2.5\mathbf{j} + c$

$c = 4\mathbf{i} + 4.5\mathbf{j}$

Hence position vector of P at time t s

$\mathbf{s} = 4\sin 3t\,\mathbf{i} + 2.5\cos 2t\,\mathbf{j} + 4\mathbf{i} + 4.5\mathbf{j}$

$= 4(\sin 3t + 1)\,\mathbf{i} + \dfrac{1}{2}(5\cos 2t + 9)\,\mathbf{j}$

(c) When $t = \dfrac{\pi}{2}$, $\mathbf{s} = 4(\sin\dfrac{3\pi}{2} + 1)\,\mathbf{i} + \dfrac{1}{2}(5\cos \pi + 9)\,\mathbf{j}$

$= 0 + (-2.5 + 4.5)\mathbf{j}$

$= 2\mathbf{j}$

Distance OP = 2 m

Exam practice ▶▶ ▶▶▶

1 An object has a mass of 3 kg and moves on a horizontal plane under the action of a constant force of $(6\mathbf{i} + 9\mathbf{j})$ N.
The position vector of the object is $\mathbf{i} - 3\mathbf{j}$ at $t = 0$ s and its velocity at this point is $(\mathbf{i} + 4\mathbf{j})$ ms^{-1}.
   (a) Find the velocity of the object when $t = 2$ s.   [3]
   (b) Calculate the distance from O when $t = 2$ s. Give your
       answer to 2 significant figures.   [4]

2 A particle moves on a horizontal plane so that at time t seconds its position vector $\mathbf{r}$ metres relative to a fixed origin O is given by $\mathbf{r} = (2t^2)\mathbf{i} + (t^2 - 2t)\mathbf{j}$.
   (a) Determine the speed of the particle when $t = 4$ s. Give your
       answer to 1 decimal place.   [5]
   (b) Show that the acceleration of the particle is constant and
       find its magnitude. Give your answer to 1 decimal place.   [3]

3 At time t s, a particle P has position vector $\mathbf{r}$ m with respect to an origin O given by:
$$\mathbf{r} = (2t - 5)\mathbf{i} + (t - 3)\mathbf{j} + (7 - 2t)\mathbf{k}.$$
(a) Show that the distance of the particle from the origin at time t s is given by
$$OP^2 = 9t^2 - 54t + 83,$$
and find the value of t when P is closest to O. [5]
(b) Find the velocity of P and determine its magnitude. [3]

(WJEC June 2005 M2 part q6)

4 A particle moves on a horizontal plane such that its velocity vector, $\mathbf{v}$ ms^{-1} at time t s is given by:
$$\mathbf{v} = 2 \cos 2t\,\mathbf{i} - \sin t\,\mathbf{j}$$
(a) Find the acceleration vector of the particle after t s. [3]
(b) When $t = 0$, the position vector of the particle is $2\mathbf{i} - \mathbf{j}$, find the position vector when $t = \dfrac{\pi}{6}$ s. [4]

5 Particle A is moving with constant velocity $-2\mathbf{i} - 2\mathbf{j} - 5\mathbf{k}$, and at time $t = 0$ s it has position vector $\mathbf{i} - 10\mathbf{k}$. Particle B is moving with constant velocity $\mathbf{i} - 8\mathbf{j} - 5\mathbf{k}$, and at time $t = 0$ s it has position vector $7\mathbf{i} + 9\mathbf{j} - 6\mathbf{k}$.
(a) Write down the position vectors of A and B at time t s. [2]
(b) Find the distance between A and B when $t = 2$ s. [3]

(WJEC June 2006 M2 q2)

14 Types of forces, resolving forces and forces in equilibrium

Prior knowledge

You will need to understand the following from your AS studies:

- Newton's laws of motion.
- Types of force.

Quick revision

Resolution of forces

Replacing a single force with two forces acting at right angles

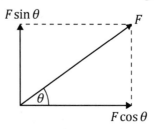

Force F can be replaced by two components at right angles to each other:

- A horizontal component $F\cos\theta$
- A vertical component $F\sin\theta$

Replacing two forces acting at right angles with a single force

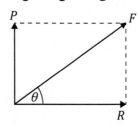

Using Pythagoras' theorem, so we obtain $F^2 = P^2 + R^2$

Hence $F = \sqrt{P^2 + R^2}$ and angle to the horizontal, $\theta = \tan^{-1}\left(\dfrac{P}{R}\right)$

The moment of a force about a point

$$\text{Moment} = \text{force} \times \text{distance}$$

The unit of moment is the newton metre (Nm) and the distance must be the perpendicular distance from the point about which moments are taken to the line of action of the force.

Moments have a sense/direction (i.e. clockwise or anticlockwise).

The principle of moments:

When a body is in equilibrium, the sum of the clockwise moments is equal to the sum of the anticlockwise moments.

Equilibrium of a rigid body under the action of parallel coplanar forces

If a rigid body, such as a rod, under the action of parallel coplanar forces is in equilibrium then the following is true:

- The resultant force in any direction is zero.

- The sum of the moments about any point is zero.

Looking at exam questions

1   Three coplanar horizontal forces of magnitude 21 N, 11 N and 8 N act on a particle P in the directions shown in the diagram:

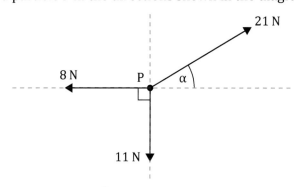

(a) Given that $\tan \alpha = \dfrac{3}{4}$, calculate the magnitude of the resultant force.   [5]

(b) Explain why the forces cannot be in equilibrium whatever the value of α.   [1]

(WJEC June 2019 Unit 4 q7)

Thinking about the question

The forces shown in the diagram probably are not in equilibrium as we are asked to find the magnitude of the resultant force. We need to find the resultant of the forces acting in the vertical direction and the resultant of the forces acting in the horizontal direction. We can use these to find the single resultant force.

For part (b) we can use a triangle of forces to show that the forces are not in equilibrium.

Starting the solution

For part (a) we find the resultant of all the horizontal forces which we will call X and then the resultant of all the vertical forces which we will call Y. We can take to the right and upwards as the positive directions for these forces. Then, by using Pythagoras' theorem, we can find the resultant of forces X and Y.

For part (b) we can construct a triangle with the two forces at right angles. We can use the two forces at right angles to calculate the third force. If this force is not equal to 21 N then the system will not form a closed triangle and the system cannot be in equilibrium.

The solution

(a) Resultant force in the horizontal direction, $X = 21\cos\alpha - 8$

Resultant force in the vertical direction, $Y = 21\sin\alpha - 11$

Now $\tan\alpha = \dfrac{3}{4}$ so $\sin\alpha = \dfrac{3}{5}$ and $\cos\alpha = \dfrac{4}{5}$

Hence, $X = 21 \times \dfrac{4}{5} - 8 = 8.8$ N and $Y = 21 \times \dfrac{3}{5} - 11 = 1.6$ N

Now, $R^2 = X^2 + Y^2$

$$= 8.8^2 + 1.6^2$$

$$R = 4\sqrt{5}\text{ N}$$

(b) For the forces to be in equilibrium, the forces must form the sides of a triangle. So the forces at right angles must have a diagonal of length equal to the other force.

Now $11^2 + 8^2 = 185$ and $21^2 = 441$

Hence, the forces cannot be in equilibrium.

> Pythagoras' theorem is used here.

2   The diagram below shows a spotlight system that consists of a symmetrical track XY that is suspended horizontally from the ceiling by means of two vertical wires:

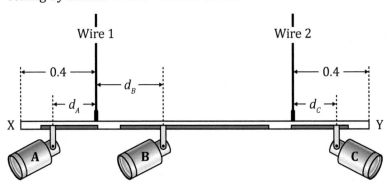

Each of the three spotlights A, B, C may be moved horizontally along its corresponding shaded section of the track. The system remains in equilibrium.

The track may be modelled as a **light** uniform rod of length 1.8 m and the wires are fixed at a distance of 0.4 m from each end. Each of the spotlights may be modelled as a particle of mass m kg, positioned at the points where they are in contact with the track.

The distances of the spotlights relative to the wires are given in the diagram and are such that:

$$0 \leq d_A \leq 0.3,\ 0.1 \leq d_B \leq 0.9,\ 0 \leq d_C \leq 0.3.$$

(a) Given that T_1 and T_2 represent the tension in wires 1 and 2 respectively, show that:

$$T_1 = mg(2 + d_A - d_B - d_C),$$

and find a similar expression for T_2.   [6]

(b) (i) Find the maximum possible value of T_1.

(ii) **Without carrying out any further calculations**, write down the maximum possible value of T_2. Give a reason for your answer.   [3]

(WJEC June 2019 Unit 4 q9)

Thinking about the question

First impressions are that this question looks quite complicated. There is a lot of information in the question and it needs reading very carefully. As the question involves forces acting at different distances, we will probably need to use the Principle of moments. Notice also that there is a range of distances that each spotlight can be moved, so this question will involve looking at the way forces change when other forces change their distances.

Starting the question

Firstly, we need to draw a diagram marking on it all the forces acting and the distances given in the question. There will be tensions in each of the strings acting upwards. These tensions would only be the same if the masses of the lamps were symmetrically positioned on the track. So, for now we will call then T_1 and T_2. The weights of the lamps (i.e. mg) are all the same. We note also that the distance between the two wires is 1 m.

For part (a) we need to take moments. Looking at the equation we are asked to find, we notice that T_2 is not in the equation. This means we need to take moments about the point where T_2 acts which we can call A. Remember that if you take moments about a point where a force acts, then the force does not have a moment, as it is zero distance to the point.

To find T_2 we need to take moments about the point where force T_1 acts which we can call B.

For part (b) we need to consider the variation of the distances of the weights of the lamps and use the inequalities in the question.

The solution

(a)

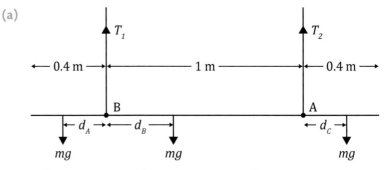

Taking moments about point A, we obtain:

$$mgd_C + T_1 \times 1 = mg(1 - d_B) + mg(1 + d_A)$$

$$mgd_C + T_1 = mg - mgd_B + mg + mgd_A$$

$$T_1 = 2mg + mgd_A - mgd_B - mgd_C$$

$$= mg(2 + d_A - d_B - d_C)$$

Taking moments about point B, we obtain:

$$mgd_B + mg(1 + d_C) = mgd_A + T_2 \times 1$$

$$mgd_B + mg + mgd_C = mgd_A + T_2$$

$$T_2 = mg(1 - d_A + d_B + d_C)$$

An alternative way of obtaining this second equation would be to resolve vertically to give $T_2 = 3mg - T_1$ and then substitute the expression for T_1 into this.

(b) (i) $T_1 = mg(2 + d_A - d_B - d_C)$

For T_1 to have its maximum value, from the above equation, d_A needs to have its maximum value and d_B and d_C need to have their minimum values.

Hence, $d_A = 0.3$, $d_B = 0.1$ and $d_C = 0$.

Max value of $T_1 = mg(2 + 0.3 - 0.1 - 0)$

$$= 2.2mg$$

(ii) As the whole arrangement is symmetrical, the maximum value of $T_2 = 2.2mg$.

Exam practice

① The diagram shows a uniform plank AD, of mass 8 kg and
length 1.0 m, resting horizontally on two supports at B and C,
where AB = 0.2 m and CD = 0.3 m. The reactions on the plank at
the supports B and C are denoted by P and Q respectively.

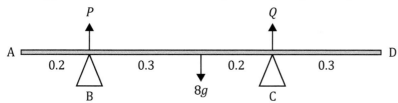

(a) Find the magnitudes of the reactions P and Q at the
supports B and C respectively. [6]
(b) Determine the greatest weight that can be placed at D
without the plank tilting. [4]

② Four coplanar forces of magnitudes 10 N, $11\sqrt{3}$ N, 16 N and 3 N
act at the point P in the directions as shown in the diagram.

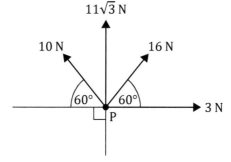

Resolve the forces in two perpendicular directions and deduce
the magnitude and direction of the resultant force. [10]

(WJEC Jan 2006 M1 q6)

③ The diagram shows a uniform straight rod AB, of length 3.8 m,
resting horizontally in equilibrium on two smooth supports at
C and D with an object of mass 2.2 kg freely suspended from
point B.

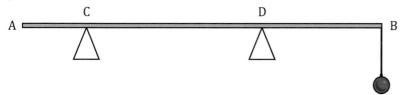

The mass of the rod is 4.4 kg, AC = 0.4 m and AD = 2.6 m.
Calculate the magnitudes of the reactions at C and D. [7]

(WJEC June 2006 q6)

4   An object, W, of weight 200 N is suspended in equilibrium by means of light inextensible strings OA, OB and OW. The strings OA and OB are inclined at 30° and 60° to the vertical respectively.

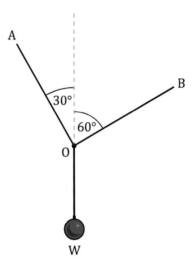

(a) Find the tension in the string OW.   [1]
(b) Draw a diagram showing the forces acting at O.   [1]
(c) Calculate the tensions in the strings OA and OB.   [8]

5

A uniform rod AB of length 1.8 m and mass 2 kg is supported by a cable AC and a support at D where distance AD = 1.6 m.
(a) Find the tension in the cable.   [3]
(b) Find the reaction at D.   [3]

6   Three horizontal forces of magnitudes 18 N, $9\sqrt{2}$ N and $8\sqrt{3}$ N act at a point in the directions shown in the diagram below:

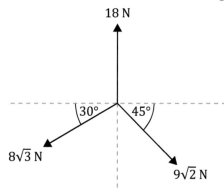

Find the magnitude of the resultant force and determine the angle it makes with the 18 N force.   [8]

(WJEC June 2019 M1 q5)

15 Forces and Newton's laws

Prior knowledge

You will need to make sure you fully understand the following from your AS studies:

- Newton's laws of motion.

- Types of force.

- The motion of particles connected by strings passing over pulleys or pegs.

Quick revision

Newton's laws of motion

1st law – a particle will remain at rest or will continue to move with constant speed in a straight line unless acted upon by some external force.

2nd law – a resultant force produces an acceleration, according to the formula

Force = mass × acceleration.

3rd law – every action has an equal and opposite reaction.

Finding a resultant force or acceleration

Resolve all the forces in two perpendicular directions and then use Pythagoras' theorem to find the resultant of these two forces at right angles. Use trigonometry to find the angle and clearly state the direction. Use $a = \dfrac{F}{m}$ to find the acceleration which will be in the direction of the resultant force.

Motion on an inclined plane

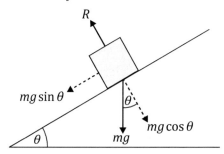

The component of the weight parallel to the slope = $mg \sin \theta$

The component of the weight at right angles to the slope = $mg \cos \theta$

Friction opposes motion

It can increase up to a certain maximum value called limiting friction or F_{MAX}.

Limiting friction, $F = \mu R$ which can also be written as $F_{MAX} = \mu R$, where μ is the coefficient of friction and R is the normal reaction between the surfaces.

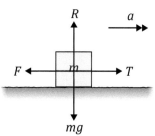

If $T > F$ there will be an unbalanced force and the mass will accelerate.

By Newton's 2nd law $\qquad ma = T - F$

Once moving, the mass will experience the maximum frictional force given by

$$F_{MAX} = \mu R$$

The forces are balanced in the vertical direction, so $R = mg$

Looking at exam questions

1

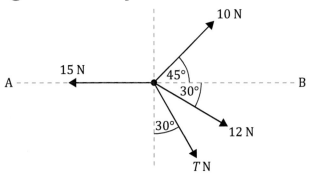

A particle of mass 0.5 kg moves along the straight line AB where it is subjected to the forces whose sizes and magnitudes are shown in the diagram.

(a) Find the magnitude of the resultant force acting. [5]

(b) Find the value of the acceleration of the particle. [2]

Thinking about the question

There is a resultant force acting but the two components at right angles to AB must be in equilibrium as there is no motion in this direction. We therefore only need to find the resultant force in the direction of AB. The acceleration can then be found from the resultant force using $F = ma$.

Starting the solution

We will resolve the forces along AB to find the resultant force and then use the equation $F = ma$ to find the value of the acceleration.

The solution

(a) The particle moves in a straight line between A and B, so this means the forces with components at right-angles to this line must be in equilibrium.

Resolving the forces at right-angles to AB, we obtain:

$$10 \sin 45° = 12 \sin 30° + T \cos 30°$$

$$10 \times \frac{1}{\sqrt{2}} = 12 \times \frac{1}{2} + T \times \frac{\sqrt{3}}{2}$$

$$T = 1.24 \text{ N}$$

Resultant force = $1.24 \cos 60° + 12 \cos 30° + 10 \cos 45° - 15$

$$= 3.08 \text{ N}$$

(b) $F = ma$

$$3.08 = 0.5a$$

$$a = 6.16 \text{ ms}^{-2}$$

---

2   A box of mass 10 kg is pulled up a rough surface, inclined at an angle of α degrees to the horizontal, by a rope parallel to the surface. The coefficient of friction between the surface and the box is 1.8 and $\sin \alpha = \frac{3}{5}$.

(a)   Draw a diagram showing all the forces acting on the box.   [4]

(b)   If the box is just at the point of slipping down the slope, show that the minimum tension in the rope needed just to prevent the box from slipping down the slope is $20.4g$ Newton.   [5]

Thinking about the question

For part (a) we need to include the following forces in the diagram: the weight, normal reaction, friction and tension in the rope. For part (b) we need to think about the forces acting parallel to the slope and how we will work out the frictional force acting down the slope.

Starting the solution

For part (a) we need to add the following to the diagram showing the box on the slope: the weight acting vertically down, the normal reaction on the box acting perpendicular to the slope, the tension in the rope acting up the slope and the frictional force acting down the slope.

For part (b) we need to consider the frictional force on the point of slipping. The formula for this, which must be recalled, is $F = \mu R$. To find the value of R it is necessary to resolve the forces at right angles to the slope.

The solution

(a)

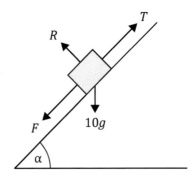

F = Frictional force

$10g$ = Weight

R = Normal reaction

T = Tension in rope

(b) As the box is just at the point of slipping, the forces on the box are in equilibrium.

Resolving forces at right-angles to the slope, we obtain

$R = 10g \cos \alpha$

On the point of slipping, $F = \mu R$

$$= 1.8 \times 10\,g \cos \alpha$$

$$= 18g \cos \alpha$$

Resolving forces parallel to the slope, we obtain

$$T = 10g \sin \alpha + 18g \cos \alpha$$

Now, as $\sin \alpha = \dfrac{3}{5}$ and $\cos \alpha = \dfrac{4}{5}$

$$T = 10g \times \frac{3}{5} + 18g \times \frac{4}{5}$$

$$= 20.4g \text{ N}$$

---

3   A box of mass 3 kg accelerates down a slope inclined at an angle of 45° to the horizontal with an acceleration of 1.5 ms⁻². Find the value of the coefficient of friction, giving your answer to 1 decimal place.   [7]

Thinking about the question

To start we need to draw a diagram adding the forces and their directions. We note that as the box accelerates down the slope, the forces in this direction are not balanced so there is a resultant force. It is important to note that the forces at right angles to the slope are balanced.

Starting the solution

There is a resultant force acting down the slope and this can be found using $F = ma$. This resultant force is the result of the component of the weight acting down the slope minus the frictional force acting up the slope. Once the frictional force has been found we can use the formula $F = \mu R$. However, we don't know the value of R, the normal reaction. To find the value of R we can use the fact that the forces perpendicular to the slope are in equilibrium. The component of the weight at right angles to the slope, $3g \cos 45°$, will equal the normal reaction, R.

The solution

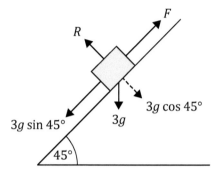

Component of the weight parallel to the slope = $3g \sin 45°$

Component of the weight at right-angles to the slope = $3g \cos 45°$

Resolving at right angles to the slope, we obtain:

$R = 3g \cos 45°$

$\quad = 3 \times 9.8 \times \dfrac{1}{\sqrt{2}}$

$\quad = 20.79 \text{ N}$

Applying Newton's 2nd law to the box, we obtain:

$\quad\quad ma = 3g \sin 45° - F$

$\quad 3 \times 1.5 = 3 \times 9.8 \times \dfrac{1}{\sqrt{2}} - F$

$\quad\quad\quad F = 16.29 \text{ N}$

$F = \mu R$, so $\mu = \dfrac{F}{R} = \dfrac{16.29}{20.79} = 0.78 = 0.8$ (1 d.p.)

> Newton's 2nd law is used as there is a resultant force which provides the acceleration down the slope.

Exam practice

1. A light inextensible string connects object A, of mass 9 kg, to object B, of mass 5 kg. The diagram shows A on a plane, inclined at an angle of 15° to the horizontal with the string passing over a smooth light pulley at the edge of the plane so that B hangs freely. The coefficient of friction between the object A and the plane is μ. Initially, A is held at rest with the string just taut. The system is then released from rest.

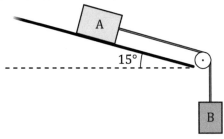

   (a) Given that $\mu = 0$, find the magnitude of the acceleration of A, and the tension in the string. [7]

   (b) Given that the object A remains at rest, find the minimum value of μ. Give your answer correct to two decimal places. [6]

   (WJEC June 2019 M1 Q6)

2. A car of mass 800 kg is travelling on a horizontal road. It experiences a resistance to motion which is constant throughout the journey. The car accelerates from rest under a constant tractive force of 300 N exerted by its engine. After 50 seconds, the car reaches a speed of 15 ms⁻¹.

   (a) Determine the magnitude of the acceleration of the car. [3]

   (b) Calculate the magnitude of the constant resistance to motion. [3]

   (c) When the car reaches the speed of 15 ms⁻¹, the engine is switched off and the car is brought to rest by a constant braking force. The total distance covered by the car for the whole journey is 500 m. Find the constant force exerted by the brakes. [7]

   (WJEC June 2017 M1 q4)

3. A sledge, of mass 39 kg, moves on a rough slope inclined at an angle α to the horizontal, where $\tan \alpha = \dfrac{5}{12}$. The coefficient of friction between the sledge and the slope is 0.3.

   (a) Given that the sledge is moving freely down a line of greatest slope, calculate the magnitude of the acceleration of the sledge. Give your answer correct to 2 decimal places. [6]

(b) Given that the sledge is being pulled up the slope with acceleration 0.4 ms⁻² by means of a rope parallel to a line of greatest slope, find the tension in the rope. [3]

(WJEC June 2012 M1 q5)

④ The diagram shows an object of mass 2 kg on a rough horizontal surface joined by a light inextensible string passing over a smooth pulley to an object of mass 5 kg, which hangs freely 1.5 m above the ground. The 5 kg mass is released from rest and takes 0.7 s to hit the ground.

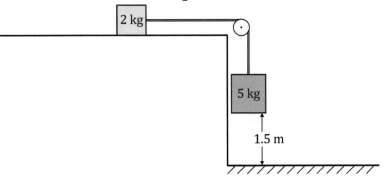

Find the coefficient of friction of the rough surface giving your answer to 1 d.p. [8]

⑤ A car of mass 1500 kg is travelling up a slope inclined at an angle α to the horizontal, where $\sin \alpha = \dfrac{1}{14}$. The resistance to motion acting on the car is 170 N. The car's engine provides a force of 3000 N.
(a) Find the acceleration of the car, giving your answer to 2 significant figures. [3]
(b) Find the coefficient of friction, giving your answer to 2 significant figures. [3]

16 Projectile motion

Prior knowledge

You will need to make sure you understand the following from your AS studies:

- Using the equations of motion (i.e. the *suvat* equations).

- Applying the *suvat* equations to vertical motion under gravity.

- Calculating the magnitude and direction of a vector and converting between component form and magnitude/direction form.

Quick revision ▶▶ ▶▶▶

The equations of motion (all of these must be memorised)

The following equations of motion are used in this topic:

$$v = u + at$$

$$s = ut + \frac{1}{2}at^2$$

$$v^2 = u^2 + 2as$$

$$s = \frac{1}{2}(u + v)t$$

s = displacement/distance
u = initial velocity/speed
v = final velocity/speed
a = acceleration
t = time

If the acceleration is zero i.e. there is constant velocity/speed,

$$\text{speed} = \frac{\text{distance travelled}}{\text{time taken}}$$

Motion of a particle projected at an angle α to the horizontal with velocity $U \, \text{m s}^{-1}$

The horizontal component, $\quad u_x = U \cos \alpha$

The vertical component, $\quad u_y = U \sin \alpha$

$$\text{Time of flight} = \frac{2U \sin \alpha}{g}$$

$$\text{Range} = \frac{U^2 \sin 2\alpha}{g}$$

$$\text{Maximum height} = \frac{U^2 \sin^2 \alpha}{2g}$$

Don't waste time trying to remember these formulae as you always need to derive them from first principles before using them.

Equation for the path of a projectile at any point (x, y) along its path is given by

$$y = x \tan \alpha - \frac{gx^2 (1 + \tan^2 \alpha)}{2U^2}$$

Looking at exam questions ▶▶ ▶▶▶

1 A ball is projected from the top edge of a vertical cliff with a velocity of $u \, \text{ms}^{-1}$ at an angle of 40° to the horizontal. The ball reaches its highest point in 2 s and then falls to the horizontal ground a vertical distance of 20 m from the top of the cliff.

(a) Calculate the value of u. [4]

(b) (i) Calculate the time of flight for the journey of the ball. [5]

(ii) Find the distance from the base of the cliff to where the ball hits the ground. [2]

Thinking about the question

As the ball is projected at an angle to the horizontal, we need to consider the vertical and horizontal components of the velocity separately with the equations of motion to answer this question. We need to be careful about the direction of the vector quantities. The displacement in the vertical direction from the top of the cliff to the point where the ball hits the ground will be –20 m if we take the upward direction as positive.

Starting the solution

For part (a) we first need to draw a diagram marking on what we know and the distances we are asked to find. We can use the equation of motion $v = u + at$ taking v at the highest point as zero, u as the vertical component of the initial velocity and a as $-g$ and t as 2 s.

For part (b)(i) we need to find the time taken to travel a vertical displacement of –20 m. Note that we need to use this negative value as this is below the point of projection. We then use $s = ut + \frac{1}{2}at^2$ to work a value for t. For part (b)(ii) we use the horizontal component of the velocity (which remains constant) along with the time of flight calculated in part (i) to work out the range using Range = horizontal component of velocity × time.

The solution

(a)

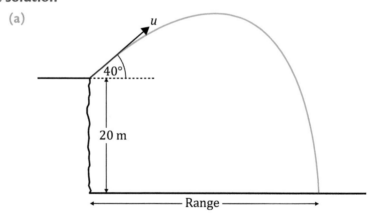

Taking the upward direction as the positive direction.

For the vertical motion, $u_v = u \sin 40°$, $v = 0$, $t = 2$, $s = -20$

Using $v = u_v + at$

$0 = u \sin 40° - 9.8 \times 2$

$u = \dfrac{19.6}{\sin 40} = 30.49 \text{ ms}^{-1}$

(b) (i) $s = ut + \frac{1}{2}at^2$

$-20 = 30.49t - 4.9t^2$

$4.9t^2 - 30.49t - 20 = 0$

$t = \frac{-b \pm \sqrt{b^2 - 4ac}}{2a} = \frac{30.49 \pm \sqrt{(30.49)^2 - 4(4.9)(-20)}}{2(4.9)}$

$t = \frac{30.49 + 36.35}{9.8}$ $s = 6.82\,s$

(ii) Range = horizontal component of velocity × time

= 30.49 cos 40° × 6.82

= 159.29 m

---

2 A tennis ball is projected with velocity vector $(30\mathbf{i} - 1.4\mathbf{j})$ ms^{-1} from a point P, which is at a height of 2.4 m vertically above a horizontal tennis court. The ball then passes over a net of height 0.9 m, before hitting the ground after $\frac{4}{7}$ s.

The unit vectors $\mathbf{i}$ and $\mathbf{j}$ are horizontal and vertical respectively. The origin O lies on the ground directly below the point P. The base of the net is x m from O.

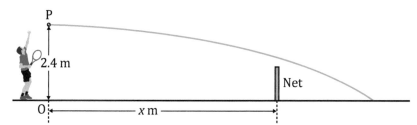

(a) Find the speed of the ball when it first hits the ground, giving your answer correct to one decimal place. [3]

(b) After $\frac{2}{5}$ s, the ball is directly above the net.

   (i) Find the position vector of the ball after $\frac{2}{5}$ s.

   (ii) Hence determine the value of x and show that the ball clears the net by approximately 16 cm. [4]

(c) In fact, the ball clears the net by only 4 cm.

   (i) Explain why the observed value is different from the value calculated in (b)(ii).

   (ii) Suggest a possible improvement to this model. [2]

   (WJEC June 2019 Unit 4 q10)

Thinking about the question

This is a question about a projectile using vectors. We need to remember that in all projectile questions, the horizontal component of the velocity remains constant whilst the vertical component is affected by gravity, which acts in the vertical direction.

Starting the solution

For part (a) we can use the equation of motion $\mathbf{v} = \mathbf{u} + \mathbf{a}t$ to find the vector of the velocity with which it hits the ground. We can then use Pythagoras' theorem to work out the speed using the horizontal and vertical components of the velocity.

For part (b)(i) we can use $\mathbf{s} = \mathbf{u}t + \frac{1}{2}\mathbf{a}t^2$ to determine the displacement vector. We need to be careful to include the vector for the height from which the ball was projected when giving the position vector.

For part (b)(ii) we can use the position vector from part (i) to obtain the distance in the horizontal direction (i.e. x) when directly over the net. To find the distance above the net we can subtract the height of the net from the number in front of the $\mathbf{j}$ in the position vector.

For part (c) we can consider the effect of air resistance which has been neglected in our answers. A differential equation could be used to take into account air resistance.

The solution

(a) $\mathbf{v} = \mathbf{u} + \mathbf{a}t$

$$= (30\mathbf{i} - 1.4\mathbf{j}) - g\mathbf{j}\left(\frac{4}{7}\right)$$

$$= 30\mathbf{i} - 7\mathbf{j}$$

Vertical velocity when it hits the ground $= -7\ \text{ms}^{-1}$ and horizontal velocity $= 30\ \text{ms}^{-1}$.

$$\text{Speed} = |\mathbf{v}| = \sqrt{(-7)^2 + 30^2}$$

$$= 30.8\ \text{ms}^{-1}\ (1\ \text{d.p.})$$

(b) (i) Using the equation of motion, $\mathbf{s} = \mathbf{u}t + \frac{1}{2}\mathbf{a}t^2$ we can write the following equation in terms of the vectors:

$$\mathbf{s} = \mathbf{u}t + \frac{1}{2}\mathbf{a}t^2$$

$$\mathbf{s} = (30\mathbf{i} - 1.4\mathbf{j})\frac{2}{5} - \frac{1}{2} \times 9.8\mathbf{j} \times \left(\frac{2}{5}\right)^2 + 2.4\mathbf{j}$$

$$= 12\mathbf{i} - 0.56\mathbf{j} - 0.784\mathbf{j} + 2.4\mathbf{j}$$

$$= 12\mathbf{i} + 1.056\mathbf{j}$$

> This equation of motion will give the displacement vector, $\mathbf{s}$, relative to the starting position which in this case is P. We require the vertical distance measured upwards from the ground, so we need to add $2.4\mathbf{j}$ to the vertical component of the displacement.

(ii) $x = 12$ m

Height of net = 0.9 m and height of ball = 1.056 m

Difference = 1.056 – 0.9 = 0.156 m = 15.6 cm

This is the multiple of **i** in the position vector in part (i).

(c) (i) The force of air resistance increases the time the ball takes to travel from P to above the net. The vertical acceleration acts for longer so the vertical distance from P to the top of the net decreases.

Another answer would be that we have ignored the radius of the ball. The larger the ball the smaller the clearance distance.

(ii) Use a differential equation where air resistance is taken into account.

Exam practice

1. A steel ball is projected with a velocity of 50 ms⁻¹ at an angle of 30° to the horizontal from the top of a vertical cliff. If the height of the cliff is 60 m and the ball hits the horizontal water below, find:
   (a) The time taken for the ball to reach the water. [4]
   (b) The horizontal distance from the base of the cliff to where the ball enters the water. [2]

2. A particle is projected from a point O with an initial velocity of 20 ms⁻¹ at an angle of θ to the horizontal where $\tan \theta = \frac{5}{12}$.
   (a) Find the vertical and horizontal components of the initial velocity. [2]
   (b) If the unit vectors acting in the horizontal and vertical directions are **i** and **j** respectively. Express the initial velocity in terms of these vectors. [2]

3. A golfer hits a ball from the point A with initial velocity 24.5 ms⁻¹ at an angle α above the horizontal, where $\sin \alpha = 0.8$. The ball just clears the tops of two trees. The tops of the trees are both 14.7 m above the level of A and are a horizontal distance d m apart.

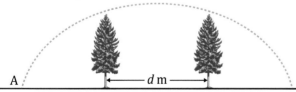

   (a) (i) Find the time taken for the ball to reach the top of the first tree.
   (ii) Determine the value of d. [8]
   (b) Find the magnitude and direction of the velocity of the ball 0.75 s after it was hit. [6]

   (WJEC June 2005 M2 q5)

4 A particle is projected from point P on the horizontal ground with a vector, **v**, given by
$$\mathbf{v} = (12\mathbf{i} + 14\mathbf{j})\ ms^{-1}.$$
The particle reaches its greatest height and then on its downward path it just clears a vertical wall which is 8.4 m tall.
(a) Find the horizontal distance of the wall from the point P. [6]
(b) Find the speed and direction of motion of the particle as it clears the wall. [7]

5 A particle is projected from a point O with an initial velocity of $(21\mathbf{i} + 12\mathbf{j})\ ms^{-1}$. The unit vectors **i** and **j** are horizontal and vertical respectively.
(a) Calculate the horizontal range of the particle. [6]
(b) Determine the maximum height reached by the particle. [3]
(c) Write down the speed and the direction of motion of the particle as it hits the ground. [1]

6 A stone is thrown from the top of a vertical cliff, 100 m above sea level. The initial velocity of the stone is 6.5 ms $^{-1}$ at an angle α degrees above the horizontal, where $\tan \alpha = \dfrac{5}{12}$.
(a) Find the time taken for the stone to reach the sea. Give your answer correct to two decimal places. [5]
(b) Calculate the horizontal distance from the bottom of the cliff to the point where the stone hits the sea. [2]
(c) Calculate the magnitude and direction of the velocity with which the stone hits the sea. [7]

(WJEC M2 June 2011 q6)

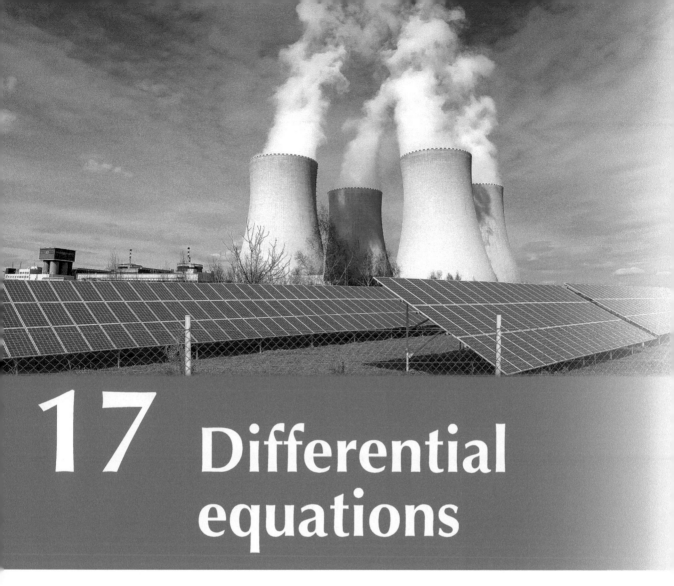

17 Differential equations

Prior knowledge

You will need to make sure you fully understand the following from your AS studies:

- Newton's laws of motion.

- Proof and use of the laws of logarithms.

- Solving equations in the form $a^x = b$.

- Integration.

Quick revision

Formation of simple differential equations

If a quantity x has a rate of **increase** in x that is proportional to x, then this can be written as a differential equation by including a constant of proportionality k as

$$\frac{dx}{dt} = kx \qquad k > 0$$

If a quantity x has a rate of **decrease** in x that is proportional to x, then this can be written as a differential equation by including a constant of proportionality k as

$$\frac{dx}{dt} = -kx \qquad k > 0$$

Separating variables and integrating

If $\quad \dfrac{dm}{dt} = -km \quad$ and $\quad m = m_0$ at $t = 0$,

then variables can be separated and integrated as follows:

$$\int \frac{1}{m}\,dm = -k\int dt$$

$$\therefore \ln m = -kt + c \tag{1}$$

When $t = 0$, $m = m_0$

Substituting these values in (1), we obtain

$$c = \ln m_0$$

Hence we obtain:

$$\ln m = -kt + \ln m_0$$

$$\ln m - \ln m_0 = -kt$$

$$\ln\left(\frac{m}{m_0}\right) = -kt$$

Taking exponentials of both sides

This is done to remove the ln from the left-hand side.

$$\frac{m}{m_0} = e^{-kt}$$

Hence $\quad m = m_0\,e^{-kt}$

Similarly, if

Usually, $f(P) = P^n$ where n is a constant.

$$\frac{dP}{dt} \propto f(P)$$

$k > 0$ if P increases with time and $k < 0$ if P decreases with time.

$$\frac{dP}{dt} = kf(P)$$

Then $\displaystyle\int \frac{1}{f(P)}\,dP = \int k\,dt$

Looking at exam questions

1 A box of mass 0.5 kg, moves along a horizontal straight line. The box is subjected to a resistive force of magnitude $3v^2$ N, where $v\,\text{ms}^{-1}$ is the speed of the box at time t seconds.

When $t = 0$, the box is at a point O and moving with speed $2\,\text{ms}^{-1}$.
$$\frac{dv}{dt} = -6v^2$$

(a) Show that v satisfies the differential equation. [2]

(b) Find an expression for v in terms of t. [4]

(c) Obtain an expression for v in terms of x, where x metres is the distance of the box from O at time t seconds. [5]

Thinking about the question

This question involves the creation of a simple differential equation which is then manipulated to form different equations. We need to be aware of the vector nature of the variables so the direction in which they act must be considered.

Starting the solution

For part (a) we notice that the only force acting is the resistive force and if we take the positive direction then this force will act in the negative direction. We can then say $m\dfrac{dv}{dt} = -$ the resistive force.

For part (b) we need to separate the variable and integrate.

For part (c) we need to turn $\dfrac{dv}{dt}$ into an expression including $\dfrac{dx}{dt}$ so we can then integrate this to obtain the expression for x.

The solution

(a) $\quad m\dfrac{dv}{dt} = -$ the resistive force

$\qquad\qquad = -3v^2$

Now $m = 0.5$ kg

$\qquad 0.5\dfrac{dv}{dt} = -3v^2$

$\qquad\quad \dfrac{dv}{dt} = -6v^2$

(b) $\frac{dv}{dt} = -6v^2$

Separating variables and integrating, we obtain:

$$\int \frac{1}{v^2}\, dv = -6\int dt$$

$$\int v^{-2}\, dv = -6\int dt$$

$$-v^{-1} = -6t + c$$

$$-\frac{1}{v} = -6t + c$$

When $t = 0$, $v = 2$

$$-2^{-1} = 0 + c$$

$$c = -\frac{1}{2}$$

$$-v^{-1} = -6t - \frac{1}{2}$$

Multiplying through by $-2v$

$$2 = 12vt + v$$

$$2 = v(12t + 1)$$

$$v = \frac{2}{12t + 1}$$

(c) $\frac{dx}{dt} = \frac{2}{12t + 1}$

We need to find $\frac{dv}{dx}$ first, so we can then separate variables and integrate.

Now $\frac{dv}{dx} = \frac{dv}{dt} \times \frac{dt}{dx}$

$$= \frac{dv}{dt} \times \frac{1}{v}$$

Now $\frac{dv}{dt} = -6v^2$, so

$$\frac{dv}{dx} = -6v^2 \times \frac{1}{v}$$

$$= -6v$$

Separating variables and integrating, we obtain:

$$\int \frac{1}{v}\,dv = -6\int dx$$

$$\ln v = -6x + c$$

When $x = 0$, $v = 2$

$$\ln 2 = 0 + c$$

$$c = \ln 2$$

Hence, $\ln v = -6x + \ln 2$

$$\ln v - \ln 2 = -6x$$

$$\ln \frac{v}{2} = -6x$$

Taking exponentials of both sides, we obtain:

$$\frac{v}{2} = e^{-6x}$$

$$v = 2e^{-6x}$$

---

2   Duckweed covers the surface of a pond and grows rapidly. The area of the pond covered by duckweed at time t days is A m². The rate of increase of A is directly proportional to A.

(a)   Write down a differential equation satisfied by A.   [2]

(b)   The area of duckweed initially is 0.5 m². After 1 week the area covered by the duckweed doubles. Find an expression for A in terms of t.   [4]

Thinking about the question

This question is about modelling duckweed growth using a differential equation. For part (a) we note that the area is increasing, so the rate of increase is positive. For part (b) we separate variables and find the constant of integration and then manipulate the equation to find the final expression.

Starting the solution

For part (a) we write an equation and introduce a constant of proportionality k. For part (b) separate the variables and integrate. We note that there will now be two constants: the constant of proportionality k and the constant of integration c. We can use the given pairs of values for A and t to find the values of k and c. We can then manipulate the equation so that it is in the form $A =$.

The solution

(a) $\dfrac{\mathrm{d}A}{\mathrm{d}t} \propto A$

$\dfrac{\mathrm{d}A}{\mathrm{d}t} = kA$

(b) $\dfrac{\mathrm{d}A}{\mathrm{d}t} = kA$

Separating variables and integrating, we obtain:

$$\int \frac{\mathrm{d}A}{A} = k\int \mathrm{d}t$$

$\ln A = kt + c$

When $t = 0$, $A = 0.5$

Hence $\ln 0.5 = 0 + c$, so $c = \ln 0.5$

We substitute this value of c back into the equation.

So, $\ln A = kt + \ln 0.5$

Watch out

t is measured in days.

Now, when $t = 7$, $A = 1$

So, $\ln 1 = 7k + \ln 0.5$

$\quad\quad 0 = 7k - 0.6931$

$\quad\quad k = 0.099$

You need to use one of the rules of logs here:

$\ln A - \ln B = \ln \dfrac{A}{B}$

Hence $\quad\quad \ln A = 0.099t + \ln 0.5$

$\quad\quad \ln A - \ln 0.5 = 0.099t$

$\quad\quad \ln\left(\dfrac{A}{0.5}\right) = 0.099t$

Taking exponentials of both sides, we obtain

$$\frac{A}{0.5} = e^{0.099t}$$

$$A = 0.5e^{0.099t}$$

Exam practice

1. A radioactive material decays in such a way so that the rate of decrease of the mass of material is proportional to the mass remaining.
   If the initial mass of radioactive material is 4 mg and 10 days later the mass is 2 mg.
   (a) Find the mass of radioactive material after 30 days. [3]
   (b) Find the time, to the nearest day, for which the mass of radioactive material has reduced to 0.3 mg. [4]

2 The mass of one of the chemicals during a chemical reaction, m, decreases at a rate which is directly proportional to m.
(a) Write down a differential equation to represent the above. [2]
(b) At time $t = 0$, the mass of the chemical is m_0. Show the equation relating mass m with mass m_0 can be modelled by the following equation: $m = m_0 e^{-kt}$ [3]
(c) (i) Sketch the graph of the equation $m = m_0 e^{-kt}$. Show on your graph, m_0. [2]
   (ii) The mass M of another chemical in the same reaction varies in such a way so that:
   $m + M = A$ where A is a constant.
   Describe what happens to the value of M as t becomes very large. [1]
   (iii) Show that $\dfrac{dM}{dt} = k(A - M)$. [2]

3 A cooling liquid obeys a law called Newton's Law of Cooling. This states that the rate of cooling is directly proportional to the difference in the temperature between the liquid temperature and the room temperature.
(a) If the temperature of the liquid is $\theta\,°C$ and the temperature of the room is $15\,°C$ and the liquid temperature is higher than the room temperature, show that:
$\ln(\theta - 15) = -kt + c$ where k and c are constants. [3]
(b) When $t = 0$ min, $\theta = 100\,°C$ and when $t = 10$ min, $\theta = 50\,°C$.
   (i) Show that $c = \ln 85$. [2]
   (ii) Hence show that $\theta = 85e^{-kt} + 15$. [4]
   (iii) Find the value of k. [3]
(c) Calculate how long it will take the liquid to reach $70\,°C$. [3]

4 The growth of bacteria can be modelled by the following differential equation:
$$\frac{dx}{dt} = \frac{kx(n - x)}{n}$$
where n and k are positive constants.

If t is the number of hours after mid-day and x is the number of bacteria after this time, show that
$$x = \frac{n}{1 + Ae^{-kt}}$$
where A is a constant. [10]

⑤ A particle moves along the x-axis such that its displacement x metres at time t seconds satisfies the differential equation:

$$\frac{dx}{dt} + x = 2$$

The particle passes through the origin when $t = 0$.

(a) Find the time when the particle reaches the point $x = 1$, and determine an expression for x at time t. [7]

(b) Hence find an expression for the acceleration of the particle at time t. [3]

(WJEC June 2017 M3 q1)

⑥ A box of mass 2 kg is projected along a horizontal surface with an initial velocity of $5\,\text{ms}^{-1}$. The box experiences a variable resistive force of $0.4v^2\,\text{N}$, where $v\,\text{ms}^{-1}$ is the velocity of the box at time t seconds.

(a) Show that v satisfies the equation:

$$5\frac{dv}{dt} + v^2 = 0$$

[3]

(b) Find an expression for v in terms of t. [4]

(c) Briefly explain why this model is not particularly realistic. [1]

(WJEC June 2019 Unit 4 q8)

Exam practice answers

Topic 1

1 Assuming there are integer values of x and y for which $9x + 12y = 1$.

$$9x + 12y = 1$$
$$3(3x + 4y) = 1$$
$$3x + 4y = \frac{1}{3}$$

Now for all integer values of x and y, $3x + 4y$ would be an integer, so not a fraction.

As this is a contradiction, the assumption is incorrect, so there are no integer values of x and y for which $9x + 12y = 1$.

> You can now see the contradiction as there is an integer value on the left and a fraction on the right.

2 It is given that $3n + 2n^3$ is odd.

Assume that n is even so that $n = 2k$.

$$3n + 2n^3 = 3(2k) + 2(2k)^3$$
$$= 6k + 16k^3$$
$$= 2(3k + 8k^3)$$

This is even as it has a factor of 2. This is a contradiction, so n is odd.

> Remember that an even number has two as a factor.

3 It is given that $2n^2 + n$ is even. We assume that n is odd.
If n is odd, it can be written as $n = 2k + 1$.

$$\text{So } 2n^2 + n = 2(2k + 1)^2 + 2k + 1$$
$$= 2(4k^2 + 4k + 1) + 2k + 1$$
$$= 8k^2 + 8k + 2 + 2k + 1$$
$$= 8k^2 + 10k + 3$$
$$= 2(4k^2 + 5k) + 3$$

Now $2(4k^2 + 5k)$ is always even as it has 2 as a factor and adding 3 to it will make the whole expression odd.
This results in $2n^2 + n$ being odd, which is a contradiction.
Hence, if $2n^2 + n$ is even, then n is even.

> $2n^2$ will always be even whether n is even or odd. Adding an odd number to it will make it odd so n must be even.

4 Assuming that there are real values of x and y such that $x^2 + 4y^2 < 4xy$.

$$x^2 + 4y^2 - 4xy < 0$$
$$x^2 - 4xy + 4y^2 < 0$$
$$(x - 2y)^2 < 0$$

But if x and y are real then $(x - 2y)^2 \geq 0$, which is a contradiction.
Hence $x^2 + 4y^2 \geq 4xy$.

5 Assuming that $\sqrt{3}$ is rational means it can be written as $\frac{a}{b}$, where a and b have no common factors.
Hence $\frac{a}{b} = \sqrt{3}$.
Squaring both sides and rearranging, we obtain:
$$a^2 = 3b^2$$

> Proving the square root of a certain number is irrational is a common exam question. Make sure you can follow these steps.

177

This means that a^2 has a factor of 3.

a has a factor 3 so that $a = 3k$, where k is an integer.

Now $a^2 = 3b^2$

$$(3k)^2 = 3b^2$$
$$9k^2 = 3b^2$$
$$3k^2 = b^2, \text{ so } b^2 \text{ has a factor 3.}$$

Hence b has a factor of 3.

This means a and b have a common factor of 3.

This is a contradiction and means that the initial assumption is wrong.

Hence $\sqrt{3}$ is irrational.

6    $$x^2 + 25 < 10x$$
$$x^2 - 10x + 25 < 0$$
$$(x - 5)(x - 5) < 0$$
$$(x - 5)^2 < 0$$

This means x cannot be real, which is a contradiction.

Hence $x + \dfrac{25}{x} \geq 10$.

7   If $x \leq -1$, let $x = -1 - a$ when $a \geq 0$.

$$(x - 3)(x + 1) \geq 0$$

So $(-1 - a - 3)(-1 - a + 1) \geq 0$
$$(-4 - a)(-a) \geq 0$$
$$a(a + 4) \geq 0$$

Now as $a \geq 0$, $a(a + 4) \geq 0$

However, $(x - 3)(x + 1) < 0$ so this is a contradiction, the assumption is false and $x > -1$.

If $x \geq 3$, let $x = 3 + b$ when $b \geq 0$.

$$(x - 3)(x + 1) \geq 0$$

So $(3 + b - 3)(3 + b + 1) \geq 0$
$$b(4 + b) \geq 0$$

Now as $b \geq 0$, $b(4 + b) \geq 0$

However, $(x - 3)(x + 1) < 0$ so this is a contradiction, the assumption is false and $x < 3$.

Hence if $(x - 3)(x + 1) < 0$ then $-1 < x < 3$.

8   Assume that $y^2 - 4x - 3 = 0$

So $y^2 = 4x + 3$

For any value of x, $4x + 3$ is always odd which means y^2 and hence y must be odd.

If y is odd, it can be written as $2n + 1$ where n is an integer.

Hence $(2n + 1)^2 = 4x + 3$
$$4n^2 + 4n + 1 = 4x + 3$$
$$4n^2 + 4n - 4x = 2$$
$$4(n^2 + n - x) = 2$$
$$n^2 + n - x = \tfrac{1}{2}$$

This is a contradiction as n and x are integers so $n^2 + n - x$ will be an integer and the left-hand side of the equation is a fraction as it is equal to $\frac{1}{2}$.

This means that the assumption that $y^2 - 4x - 3 = 0$ is incorrect so $y^2 - 4x - 3 \neq 0$ is true.

Topic 2

1 (a)

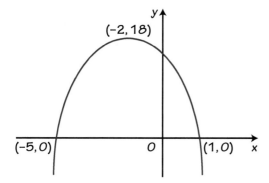

In $y = 2f(x + 3)$, the two means a stretch parallel to the y-axis with scale factor two. This means all the y-values are multiplied by 2.

The $(x + 3)$ part means a translation of $\begin{pmatrix} -3 \\ 0 \end{pmatrix}$ which moves the curve 3 units to the left.

(b)

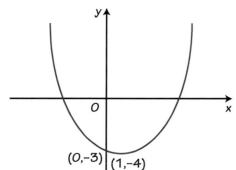

In $y = 5 - f(x)$ the $-f(x)$ means a reflection in the x-axis and the 5 means the whole curve is shifted 5 units vertically in the positive y direction.

2
$5|x + 1| = 25$
$|x + 1| = 5$
$x + 1 = \pm 5$
$x + 1 = 5$ so $x = 4$
Or $x + 1 = -5$ so $x = -6$

3
$\dfrac{2|x| + 1}{1 - |x|} = 5$
$2|x| + 1 = 5(1 - |x|)$
$2|x| + 1 = 5 - 5|x|$
$7|x| = 4$
$|x| = \dfrac{4}{7}$
$x = \pm\dfrac{4}{7}$

Remove the denominator first by multiplying both sides by it.

④ (a) $4x - 3 \geq 7$ or $4x - 3 \leq -7$
$\;\; 4x \geq 10 4x \leq -4$
$\;\; x \geq 2\frac{1}{2} x \leq -1$

(b) $\sqrt{2|x| + 2} = 4$
$\;\; 2|x| + 2 = 16$
$\;\;\; 2|x| = 14$
$\;\;\;\; |x| = 7$
$\;\;\;\;\; x = \pm 7$

Remove the square root by squaring both sides.

⑤ (a) The minus sign means we reflect the original curve in the x-axis. The number 4 means a stretch parallel to the y-axis with scale factor 4. The $(x + 3)$ means a translation of $\begin{pmatrix} -3 \\ 0 \end{pmatrix}$. Applying the reflection and stretch to $(0, -2)$ gives $(0, 8)$ and then translating this by $\begin{pmatrix} -3 \\ 0 \end{pmatrix}$ moves the point $(0, 8)$ to $(-3, 8)$. The point $(8, 0)$ stays at $(8, 0)$ after the reflection and the stretch but the translation moves the point to $(5, 0)$.

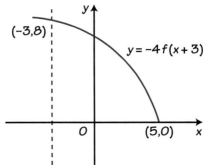

(b) The $f(2x)$ part of the function means a stretch parallel to the x-axis with scale factor $\frac{1}{2}$ and the $+ 3$ part means a translation by $\begin{pmatrix} 0 \\ 3 \end{pmatrix}$.
Applying the function, the point $(0, -2)$ will move to $(0, 1)$ and the point $(8, 0)$ will move to $(4, 3)$.

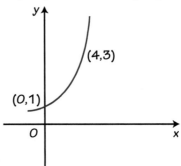

6 (a) Let
$$y = \frac{\sqrt{x^2 - 1}}{x}$$
$$xy = \sqrt{x^2 - 1}$$
$$x^2 y^2 = x^2 - 1$$
$$x^2 - x^2 y^2 = 1$$
$$x^2(1 - y^2) = 1$$
$$x^2 = \frac{1}{1 - y^2}$$
$$x = \pm \sqrt{\frac{1}{1 - y^2}}$$
$$= \frac{1}{\sqrt{1 - y^2}}$$

Watch out

If you square root you must remember to include the ± sign.

As the domain is $x \geq 1$, only the positive expression is used.
$$f^{-1} = \frac{1}{\sqrt{1 - x^2}}$$

For $f(x) = \frac{\sqrt{x^2 - 1}}{x}$, $f(x) = 0$ when $x = 1$ for the domain $x \geq 1$.
When x is very large $\sqrt{x^2 - 1} \approx \sqrt{x^2} \approx x$ and $\frac{x}{x} \to 1$ so $f(x) = 1$ is an asymptote.
$R(f) = [0, 1)$
$D(f^{-1}) = [0, 1)$

Note that the range of the function is equal to the domain of the inverse function. So $R(f) = D(f^{-1})$.

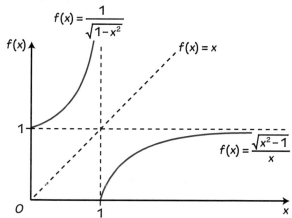

(b) $ff(x)$ cannot be formed as the range of $f(x)$ is not in the domain of $f(x)$.

7 (a) $f(x) = \sqrt{x + 5}$
When $x = -4$, $f(x) = \sqrt{5 - 4} = 1$
When $x \to \infty$, the value of $\sqrt{x + 5}$ gets larger, so $f(x) \to \infty$
$R(f) = [1, \infty)$
$g(x) = 3x^2 - 2$
When $x = 0$, $g(x) = -2$
When $x \to \infty$, the value of $3x^2 - 2$ gets larger, so $g(x) \to \infty$
$R(g) = [-2, \infty)$

(b) $gf(x) = 3(\sqrt{x+5})^2 - 2 = 13 + 3x$

(c) $fg(x) = \sqrt{3x^2 - 2 + 5} = \sqrt{3x^2 + 3}$

Now $fg(x) = 9$

$$\sqrt{3x^2 + 3} = 9$$
$$3x^2 + 3 = 81$$
$$x^2 = 26$$
$$x = \pm\sqrt{26}$$

$-\sqrt{26}$ is not in the domain of g (i.e. $[0, \infty)$) and so it is ignored. Hence $x = \sqrt{26}$

8 (a) $f(x) = x - \dfrac{1}{x}$

$f(x) = x - x^{-1}$

$f'(x) = 1 + x^{-2}$

$= 1 + \dfrac{1}{x^2}$

x^2 is positive for all positive x in the domain, so $\frac{1}{x^2}$ is also positive. Adding this to 1 will always give a positive value. Least value of $f(x)$ occurs when $x = 1$ giving a value of $f(x) = 0$.

(b) $f(x) = x - \frac{1}{x}$ for $x \geq 1$

Least value of $f(x) = 0$

When $x \to \infty$, $f(x)$ gets larger, so $f(x) \to \infty$

$R(f) = [0, \infty)$

(c) $$gf(x) = 3\left(x - \tfrac{1}{x}\right)^2 + 2$$

$$3\left(x - \tfrac{1}{x}\right)^2 + 2 = \tfrac{3}{x^2} + 8$$

$$3\left(x^2 - 2 + \tfrac{1}{x^2}\right) + 2 = \tfrac{3}{x^2} + 8$$

$$3x^2 - 6 + \tfrac{3}{x^2} + 2 = \tfrac{3}{x^2} + 8$$

$$3x^2 = 12$$

$$x = \pm 2$$

However, as $D(f)$ is $x \geq 1$ the negative value is ignored. Hence $x = 2$

Topic 3

To find the terms we need to first find the values of a and d. We form two simultaneous equations to do this and then solve them.

1 Last term = 7th term = $a + (n-1)d = a + 6d$

$a + 6d = 71$ (1)

$$S_n = \frac{n}{2}\left[2a + (n-1)d\right]$$

$$S_7 = \frac{7}{2}\left[2a + (7-1)d\right]$$

$$329 \times \frac{2}{7} = 2a + 6d$$

$$2a + 6d = 94$$

$a + 3d = 47$ (2)

Solving equations (1) and (2) simultaneously gives $d = 8$ and $a = 23$.

Hence the seven numbers are:

23, 31, 39, 47, 55, 63, 71

> Start off with a and then add d to get the next term. Add d for the second term and so on.

2 $\displaystyle\sum_{n=1}^{100}(3n - 2) = 1 + 4 + 7 + 11 + \ldots$

This is an arithmetic progression with $a = 1$, $d = 3$ and $n = 100$.

$$S_n = \frac{n}{2}\big[2a + (n-1)d\big]$$

$$S_{100} = \frac{100}{2}\big[2 + (99)(3)\big]$$

$$= 14950$$

> This formula is included in the formula booklet.

3 $\ln a + \ln c = 2\ln b$

$\ln ac = \ln b^2$

Taking exponentials of both sides, we obtain:

$$ac = b^2$$

Dividing both sides by b, we obtain:

$$\frac{ac}{b} = b$$

Dividing both sides by c, we obtain:

$$\frac{a}{b} = \frac{b}{c}$$

The common ratios of successive terms are the same, so a, b, c form a geometric sequence.

> If terms a, b, c form a geometric sequence, then the terms must have a common ratio. Hence $\frac{a}{b} = \frac{b}{c}$ so we need to prove this.

> Taking exponentials removes the ln.

4 $(a+b)^n = a^n + \dbinom{n}{1}a^{n-1}b + \dbinom{n}{2}a^{n-2}b^2 + \dbinom{n}{3}a^{n-3}b^3 + \ldots$

$(a+b)^4 = a^4 + \dbinom{4}{1}a^3 b + \dbinom{4}{2}a^2 b^2 + \dbinom{4}{3}ab^3 + \dbinom{4}{4}b^4$

> This formula is included in the formula booklet.

Finding $\dbinom{4}{1}, \dbinom{4}{2}, \dbinom{4}{3}, \dbinom{4}{4}$ by using the formula or by using Pascal's triangle and substituting them in to the above formula gives:

$$(a+b)^4 = a^4 + 4a^3 b + 6a^2 b^2 + 4ab^3 + b^4$$

$$\left(2x + \frac{1}{2x}\right)^4 = (2x)^4 + 4(2x)^3\left(\frac{1}{2x}\right) + 6(2x)^2\left(\frac{1}{2x}\right)^2 + 4(2x)\left(\frac{1}{2x}\right)^3 + \left(\frac{1}{2x}\right)^4$$

$$= 16x^4 + 16x^2 + 6 + \frac{1}{x^2} + \frac{1}{16x^4}$$

Exam practice answers

5 (a) Obtaining the formula and following the pattern in the terms gives:

$$(a + b)^n = a^n + \binom{n}{1}a^{n-1}b + \binom{n}{2}a^{n-2}b^2 + \binom{n}{3}a^{n-3}b^3 + ...$$

Substituting $n = 4$ into this formula gives:

$$(a + b)^4 = a^4 + \binom{4}{1}a^3b + \binom{4}{2}a^2b^2 + \binom{4}{3}ab^3 + \binom{4}{4}b^4$$

$$(a + b)^4 = a^4 + 4a^3b + 6a^2b^2 + 4ab^3 + b^4$$

Substituting $a = x$ and $b = \frac{2}{x}$ into the equation, gives:

$$\left(x + \frac{2}{x}\right)^4 = x^4 + 4x^3\left(\frac{2}{x}\right) + 6x^2\left(\frac{2}{x}\right)^2 + 4x\left(\frac{2}{x}\right)^3 + \left(\frac{2}{x}\right)^4$$

$$= x^4 + 8x^2 + 24 + \frac{32}{x^2} + \frac{16}{x^4}$$

(b) In the expansion of $(1 + x)^n$ the coefficient of x^2 is $\dfrac{n(n-1)}{2}$

Hence $\dfrac{n(n-1)}{2} = 55$

> Notice that this is a quadratic equation so it needs to be rearranged to equal zero so it can be factorised and solved.

$$n^2 - n = 110$$
$$n^2 - n - 110 = 0$$

Factorising this quadratic equation gives $(n - 11)(n + 10) = 0$

Solving gives $n = 11$ or -10

The question says n is a positive integer, so $n = 11$.

6 $(a + b)^n = a^n + \binom{n}{1}a^{n-1}b + \binom{n}{2}a^{n-2}b^2 + \cdots$

Here $a = a$, $b = 2x$ and $n = 5$

Putting these values into the first three terms of the formula gives:

$$(a + 2x)^5 = a^5 + \binom{5}{1}a^4(2x) + \binom{5}{2}a^3(2x)^2$$

Now $\binom{5}{1} = \dfrac{5!}{1!(5-1)!} = \dfrac{5!}{1!4!} = 5$ and $\binom{5}{2} = \dfrac{5!}{2!(5-2)!} = \dfrac{5!}{2!3!} = 10$

Hence:

$$(a + 2x)^5 = a^5 + (5)a^4(2x) + (10)a^3(2x)^2$$
$$= a^5 + 10a^4x + 40a^3x^2$$

The coefficient of x^2 is four times the coefficient of x, so

$$40a^3 = 4 \times 10a^4$$
$$40a^3 = 40a^4$$

Dividing both sides by $40a^3$ gives $a = 1$. $(a \neq 0)$

7 (a) $(1 + 4x)^{-\frac{1}{2}} = 1 + \left(-\frac{1}{2}\right)4x + \dfrac{\left(-\frac{1}{2}\right)\left(-\frac{3}{2}\right)(4x)^2}{2} + \ldots = 1 - 2x + 6x^2 + \ldots$

Expansion is valid for $|4x| < 1$ so $|x| < \dfrac{1}{4}$ or $-\dfrac{1}{4} < x < \dfrac{1}{4}$.

(b) $\quad 1 + 4y + 8y^2 = 1 + 4(y + 2y^2)$

$(1 + 4y + 8y^2)^{-\frac{1}{2}} = 1 - 2(y + 2y^2) + 6(y + 2y^2)^2 + \ldots$

$\qquad\qquad\qquad = 1 - 2y - 4y^2 + 6y^2 + \ldots$

$\qquad\qquad\qquad = 1 - 2y + 2y^2 + \ldots$

> We have to obtain $1 + 4y + 8y^2$ in the form $1 + 4x$ so we take 4 out as a factor of $4y + 8y^2$.

8 (a) The formula for the expansion of $(1 + x)^n$ is obtained from the formula booklet.

$(1 + x)^n = 1 + nx + \dfrac{n(n-1)x^2}{2!} + \dfrac{n(n-1)(n-2)x^3}{3!} + \ldots$

Putting $n = 6$ into this formula gives:

$(1 + x)^6 = 1 + 6x + \dfrac{6(5)x^2}{2!} + \dfrac{6(5)(4)x^3}{3!} + \ldots$

Note that using the first three terms only provides an approximate value.

Hence, $(1 + x)^6 \approx 1 + 6x + \dfrac{6(5)x^2}{2!} + \dfrac{6(5)(4)x^3}{3!}$

$\qquad\qquad \approx 1 + 6x + 15x^2 + 20x^3$

(b) $1 - 0.01 = 0.99$

So, $0.99^6 = (1 - 0.01)^6$

Putting $x = -0.01$ into the expansion of $(1 + x)^6$ gives:

$(1 - 0.01)^6 \approx 1 + 6(-0.01) + 15(-0.01)^2 + 20(-0.01)^3$

$\qquad\qquad \approx 0.94148$

$\qquad\qquad \approx 0.9415$ (4 decimal places)

> When obtaining a numerical answer, always check to see if the question asks that the answer needs to be given to a certain number of decimal places or significant figures. Marks can be needlessly lost not doing this.

9 $(1 - 4x)^{-\frac{1}{2}} = 1 + \left(-\frac{1}{2}\right)(-4x) + \dfrac{\left(-\frac{1}{2}\right)\left(-\frac{3}{2}\right)(4x)^2}{2} + \ldots = 1 + 2x + 6x^2 + \ldots$

Expansion is valid for $|4x| < 1$ so $|x| < \dfrac{1}{4}$.

When $x = \dfrac{1}{13}$

$(1 - 4x)^{-\frac{1}{2}} = \dfrac{1}{\sqrt{\left(1 - 4 \times \frac{1}{13}\right)}} = \dfrac{1}{\sqrt{\frac{9}{13}}} = \dfrac{\sqrt{13}}{3}$

$\dfrac{\sqrt{13}}{3} = 1 + 2\left(\dfrac{1}{13}\right) + 6\left(\dfrac{1}{13}\right)^2$

$\sqrt{13} = 3\left(1 + 2\left(\dfrac{1}{13}\right) + 6\left(\dfrac{1}{13}\right)^2\right)$

$\qquad = \dfrac{603}{169}$

(10) (a) $t_8 = ar^7 = 576$

$t_9 = ar^8 = 2304$

Now $\dfrac{ar^8}{ar^7} = r = \dfrac{2304}{576} = 4$

$ar^7 = 576$

$a(4^7) = 576$

$a = 0.03516\ldots$

$t_5 = ar^4 = 0.03516\ldots \times 4^4 = 9$

(b) (i) $ar^2 = 24$

$ar + ar^2 + ar^3 = -56$

From the first equation, $a = \dfrac{24}{r^2}$ and substituting this in for a into the second equation, we obtain:

$\left(\dfrac{24}{r^2}\right)r + \left(\dfrac{24}{r^2}\right)r^2 + \left(\dfrac{24}{r^2}\right)r^3 = -56$

$\dfrac{24}{r} + 24 + 24r = -56$

Multiplying through by r, we obtain:

$24 + 24r + 24r^2 = -56r$

$24r^2 + 80r + 24 = 0$

$3r^2 + 10r + 3 = 0$

(ii) $3r^2 + 10r + 3 = 0$

$(3r + 1)(r + 3) = 0$

$r = -\dfrac{1}{3}$ or -3

As $|r| < 1$, $r = -\dfrac{1}{3}$

Now $ar^2 = 24$ so $a(-\dfrac{1}{3})^2 = 24$ giving $a = 216$

$S_\infty = \dfrac{a}{1 - r} = \dfrac{216}{1 + \frac{1}{3}} = 162$

(11) (a) $\ln 2 + \ln (2)^2 + \ln (2)^3 + \ldots + \ln (2)^{50}$

$= \ln 2 + 2\ln 2 + 3 \ln 2 + \ldots + 50 \ln 2$

This is an arithmetic series with $a = \ln 2$ and $d = \ln 2$

$S_n = \dfrac{n}{2}\big[2a + (n - 1)d\big]$

$S_{50} = \dfrac{50}{2}\big[2 \ln 2 + (50 - 1) \ln 2\big]$

$= 25(51 \ln 2)$

$= 1275 \ln 2$

(b) $\displaystyle\sum_{n=1}^{\infty} (\ln 2)^n = \ln 2 + (\ln 2)^2 + (\ln 2)^3 + \ldots$

This is a geometric series with common ratio, $r = \ln 2$ and first term, $a = \ln 2$.

$S_\infty = \dfrac{a}{1 - r}$ provided that $|r| < 1$

$= \dfrac{\ln 2}{1 - \ln 2}$

Note that $\ln 2 < 1$, so $r < 1$.

Topic 4

1 For small angles measured in radians, we have the following identities:

$\sin x \approx x$ and $\cos x \approx 1 - \frac{1}{2}x^2$

$\sin x + \cos x = 0.5$

$x + 1 - \frac{1}{2}x^2 = 0.5$

$x^2 - 2x - 1 = 0$

$x = \dfrac{-b \pm \sqrt{b^2 - 4ac}}{2a}$

$x = \dfrac{2 \pm \sqrt{2^2 + 4}}{2}$

$= \dfrac{2 \pm \sqrt{8}}{2}$

$= \dfrac{2 \pm 2\sqrt{2}}{2}$

$= 1 - \sqrt{2}$

> Note that these are not included in the formula booklet and so need to be remembered.

> Note that the question asks for only the negative root.

2
$$3 \sin 2\theta = 2 \sin \theta$$
$$3(2 \sin \theta \cos \theta) = 2 \sin \theta$$
$$6 \sin \theta \cos \theta - 2 \sin \theta = 0$$
$$3 \sin \theta \cos \theta - \sin \theta = 0$$
$$\sin \theta (3 \cos \theta - 1) = 0$$

Hence, either $\sin \theta = 0$ or $\cos \theta = \dfrac{1}{3}$

When $\sin \theta = 0$, $\theta = 0°, 180°, 360°$

When $\cos \theta = \dfrac{1}{3}$, $\theta = 70.5°, 289.5°$

Hence $\theta = 0°, 70.5°, 180°, 289.5°, 360°$

> Use $\sin 2\theta = 2 \sin \theta \cos \theta$

> Make sure that you only find the angles in the specified range (i.e. $0° \le \theta \le 360°$)

3 $\tan 2A = \dfrac{2 \tan A}{1 - \tan^2 A}$

$\therefore \tan 2A = 3 \tan A$

gives $\dfrac{2 \tan A}{1 - \tan^2 A} = 3 \tan A$

$\therefore 2 \tan A = 3 \tan A(1 - \tan^2 A)$

$\therefore 2 \tan A = 3 \tan A - 3 \tan^3 A$

$\therefore 3 \tan^3 A - \tan A = 0$

so $\tan A(3 \tan^2 A - 1) = 0$

$\tan A = 0$ or $\tan A = \pm \dfrac{1}{\sqrt{3}}$

$A = 180°, 30°, 150°, 210°, 330°$

> Use $\tan (A + B)$ formula in the formula booklet to consider $\tan (A + A)$.

4 (a) $3 \cos \theta - 4 \sin \theta \equiv R \cos(\theta + \alpha)$
$3 \cos \theta - 4 \sin \theta \equiv R \cos \theta \cos \alpha - R \sin \theta \sin \alpha$
$3 = R \cos \alpha, 4 = R \sin \alpha$
$\therefore \tan \alpha = \frac{4}{3}, \alpha = 53.1°$ and $R = 5$

(b) $3 \cos \theta - 4 \sin \theta = 2.5$
$5 \cos(\theta + 53.1°) = 2.5$
$\cos(\theta + 53.1°) = \frac{1}{2}$
$\therefore \theta + 53.1° = 60°, 300°, 420°$
So $\theta = 6.9°, 246.9°, 366.9°$ (Note the last value is out of range and is discarded.)
Hence $\theta = 6.9°, 246.9°$

5 (a) $2 \cos^2 \theta + 6 \sin \theta \cos \theta = 2 \left(\frac{1 + \cos 2\theta}{2} \right) + 3(2 \sin \theta \cos \theta)$
$$= 1 + \cos 2\theta + 3 \sin 2\theta$$

(b) $2 \cos^2 \theta + 6 \sin \theta \cos \theta = 2$
becomes $1 + \cos 2\theta + 3 \sin 2\theta = 2$
$\therefore \cos 2\theta + 3 \sin 2\theta = 1$
Let $\cos 2\theta + 3 \sin 2\theta \equiv R \cos (2\theta - \alpha)$
$\cos 2\theta + 3 \sin 2\theta = R \cos 2\theta \cos \alpha + R \sin 2\theta \sin \alpha$
$1 = R \cos \alpha, 3 = R \sin \alpha$
$\tan \alpha = 3, \alpha = 71.57°$ and $R = \sqrt{10}$
Equation becomes $\sqrt{10} \cos(2\theta - 71.57°) = 1$
$\cos(2\theta - 71.57°) = \frac{1}{\sqrt{10}}$
$2\theta - 71.57 = -71.57°, 71.57°, 288.43°$
$\therefore 2\theta = 0°, 143.14°, 360°$
$\theta = 0°, 71.57°, 180°$

Notice that the negative angle $-71.57°$ needs to be included here.

6 Area of sector AOB $= \frac{1}{2} r^2 \theta = \frac{1}{2} r^2 (2.15) = 1.075 r^2$

Area of sector BOC $= \frac{1}{2} r^2 (\pi - 2.15) = 0.496 r^2$

Area of sector AOB − Area of sector BOC $= 1.075 r^2 - 0.496 r^2$
$$= 0.579 r^2$$

Now $0.579 r^2 = 26$
$r^2 = 44.91$
$r = 6.7$ cm

7 (a) $8 \sin \theta - 15 \cos \theta - 7 = 0$

$8 \sin \theta - 15 \cos \theta \equiv R \sin(\theta - \alpha)$

$\equiv R \sin \theta \cos \alpha - R \cos \theta \sin \alpha$

$R \cos \alpha = 8$ and $R \sin \alpha = 15$

$\tan \alpha = \dfrac{15}{8}$ so $\alpha = 61.93°$

$R = \sqrt{8^2 + 15^2} = 17$

Hence $8 \sin \theta - 15 \cos \theta = 17 \sin(\theta - 61.93°)$

(b) $17 \sin(\theta - 61.93°) = 7$

$\sin(\theta - 61.93°) = \dfrac{7}{17}$

$(\theta - 61.93°) = 24.32°, 155.68°$

Hence $\qquad \theta = 86.25°, 217.61°$

(c) $\dfrac{1}{8 \sin \theta - 15 \cos \theta + 23} = \dfrac{1}{17 \sin(\theta - 61.93°) + 23}$

Greatest value will be when the denominator has its smallest value.

Least value of $\sin(\theta - 61.93°)$ is -1.

Greatest value $= \dfrac{1}{-17 + 23} = \dfrac{1}{6}$

Least value will be when the denominator has its greatest value (i.e. when $\sin(\theta - 61.93°) = 1$)

Least value $= \dfrac{1}{17 + 23} = \dfrac{1}{40}$

8 $4 \sec^2 \theta = 7 - 11 \tan \theta$

$4(1 + \tan^2 \theta) = 7 - 11 \tan \theta$

$4 + 4 \tan^2 \theta = 7 - 11 \tan \theta$

$4 \tan^2 \theta + 11 \tan \theta - 3 = 0$

$(4 \tan \theta - 1)(\tan \theta + 3) = 0$

> Use $\sec^2 \theta = 1 + \tan^2 \theta$ to give the equation just in terms of $\tan \theta$. You have to recognise that the resulting equation is a quadratic equation.

> The tan function has a period of $180°$. Once you have found $\tan^{-1}\dfrac{1}{4}$, which is $14°$, you add $180°$ to this to find the next solution. The solutions beyond this lie outside the range.

$\tan \theta = \dfrac{1}{4}$ or $\tan \theta = -3$

When $\tan \theta = \dfrac{1}{4}$, $\theta = 14.0°$ or $194.0°$

When $\tan \theta = -3$, $\theta = 180 - 71.6° = 108.4°$

or $\theta = 360 - 71.6° = 288.4°$

Hence $\theta = 14.0°, 108.4°, 194°$

or $288.4°$ correct to one decimal place.

> tan is negative in the second and fourth quadrants, so $\theta = 180 - 71.6$ or $360 - 71.6$

> Alternatively, using the fact that the tan function has a period of $180°$, once you have found one solution, i.e. $108.4°$, you add $180°$ to find the next solution.

9 Area of sector $AOB = \dfrac{1}{2} \times r^2 \times 2 \cdot 15$

Area of sector $BOC = \dfrac{1}{2} \times r^2 \times (\pi - 2 \cdot 15)$

$\dfrac{1}{2} \times r^2 \times 2 \cdot 15 - \dfrac{1}{2} \times r^2 \times (\pi - 2 \cdot 15) = 26$

$r^2 = \dfrac{52}{4 \cdot 3 - \pi}$ (o.e.)

$r = 6 \cdot 7$

Exam practice answers

Topic 5

1 (a) $y = \ln 3x$

$y = \ln x + \ln 3$

$$\frac{dy}{dx} = \frac{1}{x}$$

Alternative solution

As $\ln 3x$ is a function of a function we can use the following method for differentiation:

Let $u = 3x$, so $\dfrac{du}{dx} = 3$

$y = \ln u$, so $\dfrac{dy}{du} = \dfrac{1}{u} = \dfrac{1}{3x}$

$\dfrac{dy}{dx} = \dfrac{dy}{du} \times \dfrac{du}{dx} = \dfrac{1}{3x} \times 3 = \dfrac{1}{x}$

(b) $y = \sqrt{1 - 3x^2}$

Let $u = 1 - 3x^2$, $\dfrac{du}{dx} = -6x$

$y = u^{\frac{1}{2}}$

$\dfrac{dy}{du} = \dfrac{1}{2}u^{-\frac{1}{2}} = \dfrac{1}{2\sqrt{1 - 3x^2}}$

$\dfrac{dy}{dx} = \dfrac{dy}{du} \times \dfrac{du}{dx} = \dfrac{1}{2\sqrt{1 - 3x^2}} \times (-6x)$

$$= \dfrac{-3x}{\sqrt{1 - 3x^2}}$$

(c) $y = x^3 \cos 3x$

$\dfrac{dy}{dx} = x^3(-3 \sin 3x) + \cos 3x \,(3x^2)$

$= -3x^3 \sin 3x + 3x^2 \cos 3x$

2 (a) Let $u = x^2$, $\dfrac{du}{dx} = 2x$

$y = \tan u$, $\dfrac{dy}{du} = \sec^2 u$

$\dfrac{dy}{dx} = \dfrac{dy}{du} \times \dfrac{du}{dx} = \sec^2 u \times 2x = 2x \sec^2 x^2$

(b) $y = e^{2x} \sin 2x$

$\dfrac{dy}{dx} = f(x)\,g'(x) + g(x)\,f'(x)$

$= e^{2x}\, 2 \cos 2x + \sin 2x\, 2e^{2x}$

$= 2e^{2x} \cos 2x + 2e^{2x} \sin 2x$

$= 2e^{2x} (\cos 2x + \sin 2x)$

3 $\dfrac{dy}{dt} = \dfrac{2t}{t^2 + 2}$

$\dfrac{dx}{dt} = 6t^2 + 2t = 2t(3t + 1)$

$\dfrac{dy}{dx} = \dfrac{dy}{dt} \times \dfrac{dt}{dx}$

$\quad = \dfrac{2t}{t^2 + 2} \times \dfrac{1}{2t(3t + 1)}$

$\quad = \dfrac{1}{(t^2 + 2)(3t + 1)}$

Here we are given two parametric equations that can be differentiated separately and then combined to give $\dfrac{dy}{dx}$.

4 $5x^3 + 2x^2 y + y^3 = 10$

Differentiating with respect to x, we obtain:

$15x^2 + 2x^2\dfrac{dy}{dx} + y(4x) + 3y^2\dfrac{dy}{dx} = 0$

$\dfrac{dy}{dx}(2x^2 + 3y^2) = -15x^2 - 4xy$

$\dfrac{dy}{dx} = \dfrac{-15x^2 - 4xy}{2x^2 + 3y^2}$

We need to differentiate this implicitly. Notice that $2x^2 y$ is a product so the Product rule must be used.

When differentiating y^3 you first differentiate with respect to y and then multiply the result by $\dfrac{dy}{dx}$.

5 Let $u = 3x^3 + 2x^2 + x - 1$

$\dfrac{du}{dx} = 9x^2 + 4x + 1$

$y = \sin u$

Note that the derivative of sin and cos are not included in the formula booklet.

$\dfrac{dy}{du} = \cos u$

$\quad = \cos(3x^3 + 2x^2 + x - 1)$

$\dfrac{dy}{dx} = \dfrac{dy}{du} \times \dfrac{du}{dx} = (9x^2 + 4x + 1)\cos(3x^3 + 2x^2 + x - 1)$

6 Let $u = 3x^2 + 4, \dfrac{du}{dx} = 6x$

$y = \ln u$

$\dfrac{dy}{du} = \dfrac{1}{u} = \dfrac{1}{3x^2 + 4}$

$\dfrac{dy}{dx} = \dfrac{dy}{du} \times \dfrac{du}{dx}$

$\quad = \dfrac{1}{3x^2 + 4} 6x$

$\quad = \dfrac{6x}{3x^2 + 4}$

There are two functions here, so this is a function of a function.

Exam practice answers

7 (a) $x^2 + 2xy + 3y^2 = 12$

Differentiating implicitly with respect to x, we obtain:

$$2x + 2x\frac{dy}{dx} + y(2) + 6y\frac{dy}{dx} = 0$$

$$2x + 2y + 2x\frac{dy}{dx} + 6y\frac{dy}{dx} = 0$$

$$2x\frac{dy}{dx} + 6y\frac{dy}{dx} = -2x - 2y$$

$$\frac{dy}{dx}(2x + 6y) = -2(x + y)$$

$$\frac{dy}{dx} = \frac{-2(x + y)}{2(x + 3y)}$$

$$= -\frac{(x + y)}{(x + 3y)}$$

> To differentiate $2xy$ we must use the Product rule.

(b) (i) $x = 2t^4$ so $\frac{dx}{dt} = 8t^3$ and $y = 3t^2$ so $\frac{dy}{dt} = 6t$

$$\frac{dy}{dx} = \frac{dy}{dt} \times \frac{dt}{dx}$$

$$= (6t)\left(\frac{1}{8t^3}\right)$$

$$= \frac{3}{4t^2}$$

(ii) $\frac{d^2y}{dx^2} = \frac{d}{dt}\left(\frac{dy}{dx}\right) \times \frac{dt}{dx}$

$$= \frac{d}{dt}\left(\frac{3}{4}t^{-2}\right) \times \frac{1}{8t^3}$$

$$= -\frac{6}{4}t^{-3} \times \frac{1}{8t^3}$$

$$= -\frac{3}{16t^6}$$

> This formula must be recalled from memory.

8 (a) Using the Product rule:

$$\frac{dy}{dx} = 12x^2 \sec^2 4x + (\tan 4x)(6x)$$

$$= 12x^2 \sec^2 4x + 6x \tan 4x$$

(b) $y = \cos^{-1} 2x$

$$\frac{dy}{dx} = -\frac{2}{\sqrt{1 - (2x)^2}} = -\frac{1}{\sqrt{1 - 4x^2}}$$

> The formulae for the derivatives of the inverse trig functions are included in the formula booklet.
>
> Note that the formula for the derivative of $\cos^{-1} x$ is $-\dfrac{1}{\sqrt{1 - x^2}}$.
>
> Here we simply replace the x in the above formula with $2x$.

(c) $y = \dfrac{\ln x}{x^3}$

$\dfrac{dy}{dx} = \dfrac{f'(x)\,g(x) + f(x)\,g'(x)}{(g(x))^2}$

$\dfrac{dy}{dx} = \dfrac{\left(\frac{1}{x}\right)(x^3) - (\ln x)(3x^2)}{(x^3)^2}$

$= \dfrac{x^2 - 3x^2 \ln x}{x^6}$

$= \dfrac{x^2(1 - 3 \ln x)}{x^6}$

$= \dfrac{1 - 3 \ln x}{x^4}$

9 $\dfrac{dr}{dt} = 1.5$

$\dfrac{dA}{dt} = \dfrac{dr}{dt} \times \dfrac{dA}{dr}$

Look at the question carefully to spot any rates of change.

Now $A = \pi r^2$, so $\dfrac{dA}{dr} = 2\pi r$

Hence,

$\dfrac{dA}{dt} = \dfrac{dr}{dt} \times \dfrac{dA}{dr}$

$= 1.5 \times 2\pi r$

When $r = 100$, $\dfrac{dA}{dt} = 1.5 \times 2\pi \times 100 = 942 \text{ m}^2 \text{ min}^{-1}$

In this type of question you often need to find an equation to differentiate. Usually they are for working out areas or volumes.

10 $y = \sin^{-1}\left(\dfrac{1}{x}\right)$

Let $u = \dfrac{1}{x}$ so $u = x^{-1}$ and $\dfrac{du}{dx} = -x^{-2} = -\dfrac{1}{x^2}$

This is a function of a function, so we let u equal the contents of the bracket.

$y = \sin^{-1} u$ so $\dfrac{dy}{du} = \dfrac{1}{\sqrt{1 - u^2}}$

$\dfrac{dy}{dx} = \dfrac{dy}{du} \times \dfrac{du}{dx} = \dfrac{1}{\sqrt{1 - u^2}}\left(-\dfrac{1}{x^2}\right)$

$= \dfrac{1}{\sqrt{1 - \left(\frac{1}{x}\right)^2}} \times \left(-\dfrac{1}{x^2}\right)$

$= -\dfrac{1}{x^2\sqrt{1 - \dfrac{1}{x^2}}}$

$= -\dfrac{1}{x^2\sqrt{\dfrac{x^2 - 1}{x^2}}}$

$= -\dfrac{1}{x\sqrt{x^2 - 1}}$

The derivative of $\sin^{-1} x$ is included in the formula booklet.

(11) Note that as this is a product, we will use the Product rule.

$$\frac{dy}{dx} = (x^2 + 1)\left(\frac{1}{x^2 + 1}\right) + (\tan^{-1}x)(2x)$$

$$= 1 + 2x\tan^{-1}x$$

(12) $y = \cos^{-1}(3x^2)$

Let $u = 3x^2$ so $\frac{du}{dx} = 6x$

$y = \cos^{-1}u$ so $\frac{dy}{du} = -\frac{1}{\sqrt{1 - u^2}} = -\frac{1}{\sqrt{1 - 9x^4}}$

$$\frac{dy}{dx} = \frac{dy}{du} \times \frac{du}{dx} = -\frac{1}{\sqrt{1 - 9x^4}}(6x)$$

$$= -\frac{6x}{\sqrt{1 - 9x^4}}$$

(13) $x^3 + 4xy - 2x - y^3 + 1 = 0$

Differentiating implicitly with respect to x, we obtain:

$$3x^2 + 4x\frac{dy}{dx} + 4y - 2 - 3y^2\frac{dy}{dx} = 0$$

$$4x\frac{dy}{dx} - 3y^2\frac{dy}{dx} = 2 - 3x^2 - 4y$$

$$\frac{dy}{dx}(4x - 3y^2) = 2 - 3x^2 - 4y$$

$$\frac{dy}{dx} = \frac{2 - 3x^2 - 4y}{4x - 3y^2}$$

At P(1, 2), $\frac{dy}{dx} = \frac{2 - 3(1)^2 - 4(2)}{4(1) - 3(2)^2} = \frac{-9}{-8} = \frac{9}{8}$

Equation of tangent at P

> Use the formula for a straight line
>
> $y - y_1 = m(x - x_1)$

$$y - 2 = \frac{9}{8}(x - 1)$$

$$8y - 16 = 9x - 9$$

$$8y - 9x - 7 = 0$$

(14) $y = x^4 - 6x^2 + 2$

$$\frac{dy}{dx} = 4x^3 - 12x$$

$$\frac{d^2y}{dx^2} = 12x^2 - 12$$

> Need to check the sign of $\frac{d^2y}{dx^2}$ either side of these two x-coordinates to check that there is a sign change. The sign change indicates they are points of inflection.

$$\frac{d^2y}{dx^2} = 0, \text{ so } 12x^2 - 12 = 0$$

$$x^2 - 1 = 0$$

$$(x + 1)(x - 1) = 0$$

$$x = 1 \text{ or } -1$$

When $x = \dfrac{1}{2}, \dfrac{d^2y}{dx^2} = 12\left(\dfrac{1}{2}\right)^2 - 12 = -9$

When $x = \dfrac{3}{2}, \dfrac{d^2y}{dx^2} = 12\left(\dfrac{3}{2}\right)^2 - 12 = 15$

There is a change in sign so at $x = 1$, there is a point of inflection.

When $x = -\dfrac{1}{2}, \dfrac{d^2y}{dx^2} = 12\left(-\dfrac{1}{2}\right)^2 - 12 = -9$

When $x = -\dfrac{3}{2}, \dfrac{d^2y}{dx^2} = 12\left(-\dfrac{3}{2}\right)^2 - 12 = 15$

There is a change in sign so at $x = -1$, there is a point of inflection.

When $x = 1, y = 1^4 - 6(1)^2 + 2 = -3$

When $x = -1, y = (-1)^4 - 6(-1)^2 + 2 = -3$

Coordinates are $(1, -3)$ and $(-1, -3)$

Topic 6

1. $x = 5t + 1$, so $t = \dfrac{x - 1}{5}$

   Now $y = 1 - 2t$ so $y = 1 - 2\left(\dfrac{x - 1}{5}\right)$

   $y = \dfrac{5 - 2(x - 1)}{5}$

   $y = \dfrac{7 - 2x}{5}$

2. $3y^2 + 6xy - 3x^2 = 4$

   Differentiating implicitly with respect to x, we obtain:

   $6y\dfrac{dy}{dx} + 6x\dfrac{dy}{dx} + y(6) - 6x = 0$

   $6y\dfrac{dy}{dx} + 6x\dfrac{dy}{dx} = 6x - 6y$

   $y\dfrac{dy}{dx} + x\dfrac{dy}{dx} = x - y$

   $\dfrac{dy}{dx}(x + y) = x - y$

   $\dfrac{dy}{dx} = \dfrac{x - y}{x + y}$

   > Notice we can simplify by dividing both sides by 6.

3. (a) $x = 5t^2$

   $t^2 = \dfrac{x}{5}$

   Now $y = 2t^4$ so $y = 2\left(\dfrac{x}{5}\right)^2 = \dfrac{2x^2}{25}$

Exam practice answers

(b) (i) $\dfrac{dx}{dt} = 10t$ and $\dfrac{dy}{dt} = 8t^3$

$$\dfrac{dy}{dx} = \dfrac{dy}{dt} \times \dfrac{dt}{dx}$$

$$= 8t^3 \times \dfrac{1}{10t}$$

$$= \dfrac{4}{5}t^2$$

At P, $t = p$ so gradient at $P = \dfrac{4}{5}p^2$

(ii) $\quad y - y_1 = m(x - x_1).$

$$y - 2p^4 = \dfrac{4}{5}p^2\,(x - 5p^2)$$

$$5y - 10p^4 = 4p^2\,(x - 5p^2)$$

$$5y = 4p^2 x - 10p^4$$

$$y = \dfrac{4}{5}p^2 x - 2p^4$$

④ $x = 2\cos 3t$

$\dfrac{dx}{dt} = -6\sin 3t$

$y = 2\sin 3t$

$\dfrac{dy}{dt} = 6\cos 3t$

$\dfrac{dy}{dx} = \dfrac{dy}{dt} \times \dfrac{dt}{dx}$

$$= \dfrac{6\cos 3t}{-6\sin 3t}$$

$$= -\dfrac{\cos 3t}{\sin 3t}$$

P is the point $(2\cos 3p, 2\sin 3p)$

Gradient of tangent at $P = -\dfrac{\cos 3p}{\sin 3p}$

Equation of tangent at P is

$$y - 2\sin 3p = -\dfrac{\cos 3p}{\sin 3p}(x - 2\cos 3p)$$

$$y\sin 3p - 2\sin^2 3p = -x\cos 3p + 2\cos^2 3p$$

$$y\sin 3p + x\cos 3p = 2\sin^2 3p + 2\cos^2 3p$$

$$y\sin 3p + x\cos 3p = 2(\sin^2 3p + \cos^2 3p)$$

$$y\sin 3p + x\cos 3p = 2$$

$$y\sin 3p + x\cos 3p - 2 = 0$$

(b) When $y = 0$ and $p = \dfrac{\pi}{3}$

$$0 + x\cos\left(\dfrac{3\pi}{3}\right) - 2 = 0$$

$$-x - 2 = 0$$

$$x = -2$$

Hence, A is the point $(-2, 0)$

5 (a) $x = \sin \theta$, so $\dfrac{dx}{d\theta} = \cos \theta$

$y = \cos 2\theta$, so $\dfrac{dy}{d\theta} = -2 \sin 2\theta$

$\dfrac{dy}{dx} = \dfrac{dy}{d\theta} \times \dfrac{d\theta}{dx} = \dfrac{-2 \sin 2\theta}{\cos \theta} = \dfrac{-4 \sin \theta \cos \theta}{\cos \theta} = -4 \sin \theta$

When $\theta = \dfrac{\pi}{4}$, $\dfrac{dy}{dx} = -4 \sin\dfrac{\pi}{4} = -4 \times \dfrac{1}{\sqrt{2}} = -\dfrac{4}{\sqrt{2}} \times \dfrac{\sqrt{2}}{\sqrt{2}} = -2\sqrt{2}$

Equation of tangent is:

$y - \cos\left(\dfrac{2\pi}{4}\right) = -2\sqrt{2}\left(x - \sin\left(\dfrac{\pi}{4}\right)\right)$

$y - \cos\left(\dfrac{\pi}{2}\right) = -2\sqrt{2}\left(x - \sin\left(\dfrac{\pi}{4}\right)\right)$

$y - 0 = -2\sqrt{2}\left(x - \dfrac{1}{\sqrt{2}}\right)$

$y = -2\sqrt{2}x + 2$

Comparing this equation with $y = mx + c$, $m = -2\sqrt{2}$ and $c = 2$.

(b) $x + y = 1$

$\sin \theta + \cos 2\theta = 1$

Now $\cos 2\theta = 1 - 2 \sin^2 \theta$

$\sin \theta + 1 - 2 \sin^2 \theta = 1$

$\sin \theta - 2 \sin^2 \theta = 0$

Now $\sin \theta = x$, so $x - 2x^2 = 0$

$x(1 - 2x) = 0$, hence $x = 0$ or $x = \dfrac{1}{2}$

Using $x + y = 1$, when $x = 0$, $y = 1$ and when $x = \dfrac{1}{2}$, $y = \dfrac{1}{2}$.

Hence, coordinates are $(0, 1)$ and $(\dfrac{1}{2}, \dfrac{1}{2})$

> This trig identity should be remembered although it can be worked out using other trig identities included in the formula booklet.

Topic 7

1 (a) $\displaystyle\int e^{6 - 4x}\, dx = \dfrac{e^{6 - 4x}}{-4}$

$= -\dfrac{1}{4}(e^{6 - 4x}) + c$

(b) $\displaystyle\int x^3 \ln x\, dx$

Obtaining the formula for integration by parts from the formula booklet, we have:

$\displaystyle\int u\dfrac{dv}{dx}\, dx = uv - \int v\dfrac{du}{dx}\, dx$

As $\ln x$ is hard to integrate, we need to make this equal to u in the formula.

So, $u = \ln x$ and $\dfrac{dv}{dx} = x^3$

Exam practice answers

Substituting these values into the formula, we obtain:

$$\int x^3 \ln x \, dx = \ln x \cdot \frac{x^4}{4} - \int \frac{x^4}{4} \cdot \frac{1}{x} \, dx$$

$$= \frac{x^4}{4} \ln x - \int \frac{x^3}{4} \, dx$$

$$= \frac{x^4}{4} \ln x - \frac{x^4}{16} + C$$

2 (a) $\dfrac{4x}{(x+1)(x-3)^2} \equiv \dfrac{A}{x+1} + \dfrac{B}{x-3} + \dfrac{C}{(x-3)^2}$

$4x \equiv A(x-3)^2 + B(x-3)(x+1) + C(x+1)$

Let $x = 3$, $12 = 0 + 0 + 4C$ giving $C = 3$

Let $x = -1$, $-4 = 16A$ giving $A = -\dfrac{1}{4}$

Let $x = 0$, $0 = 9A - 3B + C$ so $0 = -\dfrac{9}{4} - 3B + 3$ giving $B = \dfrac{1}{4}$

The partial fractions are $-\dfrac{1}{4(x+1)} + \dfrac{1}{4(x-3)} + \dfrac{3}{(x-3)^2}$

(b) $\displaystyle\int_4^5 \frac{4x}{(x+1)(x-3)^2} \, dx = \int_4^5 \left(-\frac{1}{4(x+1)} + \frac{1}{4(x-3)} + \frac{3}{(x-3)^2} \right) dx$

$$= \frac{1}{4} \int_4^5 \left(-\frac{1}{x+1} + \frac{1}{x-3} + \frac{12}{(x-3)^2} \right) dx$$

$$= \frac{1}{4} \int_4^5 \left(-\frac{1}{x+1} + \frac{1}{x-3} + 12(x-3)^{-2} \right) dx$$

$$= \frac{1}{4} \left[-\ln(x+1) + \ln(x-3) + \frac{12}{(-1)}(x-3)^{-1} \right]_4^5$$

$$= \frac{1}{4} \left[\ln\frac{(x-3)}{(x+1)} - \frac{12}{x-3} \right]_4^5$$

$$= \frac{1}{4} \left[\left(\ln\frac{1}{3} - 6 \right) - \left(\ln\frac{1}{5} - 12 \right) \right]$$

$$= \frac{1}{4} \left[\ln\frac{5}{3} + 6 \right]$$

$$= 1.63 \text{ (2 d.p.)}$$

We have taken $\dfrac{1}{4}$ out as a factor to simplify the denominators of the fractions before integration.

3 $\dfrac{dy}{dx} = 4x^3 y$

Separating variables and integrating, we obtain:

$$\int \frac{1}{y} \, dy = \int 4x^3 \, dx$$

$$\ln y = \frac{4x^4}{4} + c$$

$$\ln y = x^4 + c$$

When $x = 2$, $y = 1$ so $\ln 1 = 2^4 + c$

$0 = 16 + c$ giving $c = -16$

So, $\ln y = x^4 - 16$

Taking exponentials of both sides

$y = e^{x^4 - 16}$

Note that the question asks for the answer to be given in the form $y = f(x)$. We therefore remove the ln function by taking exponentials of both sides.

4 (a) $\int_0^1 xe^{-3x}\,dx$

$\int u\dfrac{dv}{dx}\,dx = uv - \int v\dfrac{du}{dx}\,dx$

Let $u = x$ and $\dfrac{dv}{dx} = e^{-3x}$

$\int_0^1 xe^{-3x}\,dx = \left[x\left(-\dfrac{1}{3}e^{-3x}\right)\right]_0^1 - \int_0^1 \dfrac{1}{3}e^{-3x}(1)\,dx$

$= \left[x\left(-\dfrac{1}{3}e^{-3x}\right)\right]_0^1 - \left[-\dfrac{1}{3}e^{-3x}\right]_0^1$

$= \left[-\dfrac{1}{3}xe^{-3x} + \dfrac{1}{3}e^{-3x}\right]_0^1$

$= \left[\left(-\dfrac{1}{3}e^{-3} + \dfrac{1}{3}e^{-3}\right) - \left(0 + \dfrac{1}{3}\right)\right]$

$= -\dfrac{1}{3}$

(b) $\int_0^1 \sqrt{(1-x^2)}\,dx$

Let $x = \sin\theta$ $\qquad \dfrac{dx}{d\theta} = \cos\theta$ so $dx = \cos\theta\,d\theta$

When $x = 1$, $\sin\theta = 1$ so $\theta = \dfrac{\pi}{2}$

When $x = 0$, $\sin\theta = 0$ so $\theta = 0$

Integral $= \int_0^{\frac{\pi}{2}} \sqrt{(1-\sin^2\theta)}\cos\theta\,d\theta$

Now $1 - \sin^2\theta = \cos^2\theta$ so $\sqrt{(1-\sin^2\theta)} = \sqrt{\cos^2\theta} = \cos\theta$

Integral $= \int_0^{\frac{\pi}{2}} \cos\theta\cos\theta\,d\theta$

$= \int_0^{\frac{\pi}{2}} \cos^2\theta\,d\theta$

Now $\cos^2\theta = 1 + \cos 2\theta$

Integral $= \int_0^{\frac{\pi}{2}} (1 + \cos 2\theta)\,d\theta$

$= \left[\theta + \dfrac{\sin 2\theta}{2}\right]_0^{\frac{\pi}{2}}$

$= \left[\left(\dfrac{\pi}{2} + \dfrac{\sin\pi}{2}\right) - \left(0 + \dfrac{\sin 0}{2}\right)\right]$

$= \dfrac{\pi}{2}$

5 (a) (i) $\int \sin(1-2x)\,dx = \dfrac{-\cos(1-2x)}{-2} + c$

$= \dfrac{1}{2}\cos(1-2x) + c$

(ii) $\int \dfrac{3}{e^{3x-1}}\,dx = \int 3e^{-(3x-1)}\,dx$

$= \int 3e^{1-3x}\,dx$

$= \dfrac{3e^{1-3x}}{-3} + c$

$= -e^{1-3x} + c$

Here we have the product of two functions to integrate, so we need to integrate using integration by parts. The formula is looked up in the formula booklet.

Note that it is easy to integrate both x and e^{-3x}. It is best to let u be the function that gives the simplest differential. So we choose $u = x$ in this case.

The exponential power stays the same but we need to divide by its derivative to obtain the integral.

Exam practice answers

(iii) $\int \dfrac{3}{\frac{1}{2}x - 1}\, dx = \dfrac{1}{\frac{1}{2}} \int \dfrac{(3)\left(\frac{1}{2}\right)}{\frac{1}{2}x - 1}\, dx$

$$= 6 \int \dfrac{\frac{1}{2}}{\frac{1}{2}x - 1}\, dx$$

$$= 6 \ln\left(\dfrac{1}{2}x - 1\right) + c$$

(b) $\displaystyle\int_1^5 \sqrt{(2x - 1)}\, dx = \int_1^5 (2x - 1)^{+\frac{1}{2}}\, dx$

$$= \left[\dfrac{(2x - 1)^{\frac{1}{2}}}{\left(\frac{1}{2}\right)(2)}\right]_1^5$$

$$= \left[\sqrt{(2x - 1)}\right]_1^5$$

$$= 3 - 1$$

$$= 2$$

6 Let $x = \sin\theta$ $\dfrac{dx}{d\theta} = \cos\theta$

$dx = \cos\theta\, d\theta$

Now when $x = 1$, $\sin\theta = 1$ giving $\theta = \dfrac{\pi}{2}$

when $x = 0$, $\sin\theta = 0$ giving $\theta = 0$

Integral $= \displaystyle\int_0^{\frac{\pi}{2}} \dfrac{8 \sin^2\theta}{\sqrt{(1 - \sin^2\theta)}} \cos\theta\, d\theta$

$$= \int_0^{\frac{\pi}{2}} \dfrac{8 \sin^2\theta}{\sqrt{(\cos^2\theta)}} \cos\theta\, d\theta$$

$$= \int_0^{\frac{\pi}{2}} \dfrac{8 \sin^2\theta}{\cos\theta} \cos\theta\, d\theta$$

$$= \int_0^{\frac{\pi}{2}} 8 \sin^2\theta\, d\theta$$

$$= 8 \int_0^{\frac{\pi}{2}} \dfrac{1 - \cos 2\theta}{2}\, d\theta$$

$$= 4 \int_0^{\frac{\pi}{2}} (1 - \cos 2\theta)\, d\theta$$

$$= 4 \left[\theta - \dfrac{\sin 2\theta}{2}\right]_0^{\frac{\pi}{2}}$$

$$= 4 \left[\left(\dfrac{\pi}{2} - \dfrac{\sin \pi}{2}\right) - \left(0 - \dfrac{\sin 0}{2}\right)\right]$$

$$= 2\pi$$

7 Let $u = 1 + x^3$, so $\frac{du}{dx} = 3x^2$ and $dx = \frac{du}{3x^2}$

Integral now becomes $\int 6x^2 u^{\frac{1}{2}} \frac{du}{3x^2}$

$$= 2\int u^{\frac{1}{2}} du$$

$$= 2\frac{u^{\frac{3}{2}}}{\frac{3}{2}} + c$$

$$= 2 \times \frac{2}{3} u^{\frac{3}{2}} + c$$

$$= \frac{4}{3} u^{\frac{3}{2}} + c$$

$$= \frac{4}{3} \sqrt{(1 + x^3)^3} + c$$

Divide top and bottom by $3x^2$.

8 (a) $\int \frac{4x^3 + 3x^2}{x^4 + x^3} dx = \ln(x^4 + x^3) + c$

(b) $\int \frac{\cos x}{\sin^2 x} dx$

Let $u = \sin x$ so $\frac{du}{dx} = \cos x$ and $dx = \frac{du}{\cos x}$

Integral becomes $\int \frac{\cos x}{\sin^2 x} \frac{du}{\cos x}$

$$= \int \frac{1}{u^2} du$$

$$= \int u^{-2} du$$

$$= \frac{u^{-1}}{-1} + c$$

$$= -\frac{1}{u} + c$$

$$= -\frac{1}{\sin x} + c$$

Always check when you are integrating a fraction that the top can be made into the derivative of the bottom. If this is the case, then the answer will involve ln.

9 $\frac{x + 1}{(x - 1)(x - 2)(x - 3)} = \frac{A}{x - 1} + \frac{B}{x - 2} + \frac{C}{x + 3}$

$x + 1 = A(x - 2)(x - 3) + B(x - 1)(x - 3) + C(x - 1)(x - 2)$
Let $x = 1$, $2 = 2A$, $A = 1$
Let $x = 2$, $3 = -B$, $B = -3$
Let $x = 3$, $4 = 2C$, $C = 2$

$$\int_4^5 \frac{x + 1}{(x - 1)(x - 2)(x - 3)} dx = \int_4^5 \left(\frac{1}{x - 1} - \frac{3}{x - 2} + \frac{2}{x - 3}\right) dx$$

$$= \left[\ln(x - 1) - 3\ln(x - 2) + 2\ln(x - 3)\right]_4^5$$

$$= [(\ln 4 - 3\ln 3 + 2\ln 2) - (\ln 3 - 3\ln 2 + 2\ln 1)]$$

$$= [\ln 4 - 3\ln 3 + 2\ln 2 - \ln 3 + 3\ln 2 - 2\ln 1]$$

$$= \ln 4 - 4\ln 3 + 5\ln 2 + 0$$

$$= \ln 4 - \ln 81 + \ln 32$$

$$= \ln\left(\frac{128}{81}\right)$$

(10) (a) $\dfrac{\mathrm{d}A}{\mathrm{d}t} \propto \sqrt{A}$

$\dfrac{\mathrm{d}A}{\mathrm{d}t} = k\sqrt{A}$

(b) Separating variables and integrating, we obtain:

$$\int \dfrac{\mathrm{d}A}{\sqrt{A}} = \int k\,\mathrm{d}t$$

$$\int A^{-\frac{1}{2}}\,\mathrm{d}A = \int k\,\mathrm{d}t$$

$$\dfrac{A^{\frac{1}{2}}}{\frac{1}{2}} = kt + c$$

$$2\sqrt{A} = kt + c$$

When $A = 64$, $t = 3$ so $2\sqrt{64} = 3k + c$

$16 = 3k + c$

When $A = 196$, $t = 5.5$ so $2\sqrt{196} = 5.5k + c$

$28 = 5.5k + c$

Solving these two equations simultaneously gives $k = 4.8$ and $c = 1.6$

Substituting these values into the equation $2\sqrt{A} = kt + c$ we obtain: $2\sqrt{A} = 4.8t + 1.6$

$\sqrt{A} = 2.4t + 0.8$

Squaring both sides to remove the square root, we obtain:

$$A = (2.4t + 0.8)^2$$

> Notice that we can divide both sides of this equation by 2.

(11) (a) $\dfrac{\mathrm{d}A}{\mathrm{d}t} \propto A$

$\dfrac{\mathrm{d}A}{\mathrm{d}t} = kA$

(b) Separating variables and integrating, we obtain:

$$\int \dfrac{\mathrm{d}A}{A} = k\int \mathrm{d}t$$

$\ln A = kt + c$

When $t = 0$, $A = 0.2$, so $\ln 0.2 = 0 + c$, giving $c = \ln 0.2$.

Hence $\ln A = kt + \ln 0.2$

$\ln A - \ln 0.2 = kt$

$\ln 5A = kt$

$5A = \mathrm{e}^{kt}$

The answer needs to be in the form pq^t, where p and q are both rational. This means we need to find an value for e^k so that e is not included in the final expression.

When $t = 1$, $A = 1.48$ so $\ln 7.4 = k$

$\mathrm{e}^k = 7.4$

Now substitute this into the previous expression and rearrange.

$5A = \mathrm{e}^{kt}$

$A = 0.2(7.4)^t$

⑫ $\displaystyle\int_0^{\frac{\pi}{3}} 2\cos\left(3x + \frac{\pi}{3}\right) dx$

$= 2\displaystyle\int_0^{\frac{\pi}{3}} \cos\left(3x + \frac{\pi}{3}\right) dx$

$= 2\left[\dfrac{\sin\left(3x + \frac{\pi}{3}\right)}{3}\right]_0^{\frac{\pi}{3}}$

$= \dfrac{2}{3}\left[\sin\left(3x + \frac{\pi}{3}\right)\right]_0^{\frac{\pi}{3}}$

$= \dfrac{2}{3}\left[\sin\left(3\frac{\pi}{3} + \frac{\pi}{3}\right) - \left(\sin\left(3(0) + \frac{\pi}{3}\right)\right)\right]$

$= \dfrac{2}{3}\left[\sin\dfrac{4\pi}{3} - \sin\dfrac{\pi}{3}\right]$

$= \dfrac{2}{3}\left(-\dfrac{\sqrt{3}}{2} + \dfrac{\sqrt{3}}{2}\right)$

$= 0$

Topic 8

① (a)

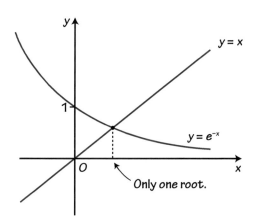

Only one root.

(b) $e^{-x} = x$

$f(x) = x - e^{-x}$

$f'(x) = 1 + e^{-x}$

$x_{n+1} = x_n - \dfrac{f(x_n)}{f'(x_n)}$

$x_1 = x_0 - \dfrac{f(x_0)}{f'(x_0)} = 0.6 - \dfrac{f(0.6)}{f'(0.6)} = 0.6 - \dfrac{(0.6 - e^{-0.6})}{(1 + e^{-0.6})} = 0.5669$

$x_2 = 0.5671$

$x_3 = 0.5671$

Hence solution is $x = 0.567$ correct to 3 d.p.

Exam practice answers

Remember that we need to be working in radians.

2 $\sin x + x - 1 = 0$

$\sin x = -x + 1$

We now need to sketch the graphs $y = \sin x$ and $y = -x + 1$.

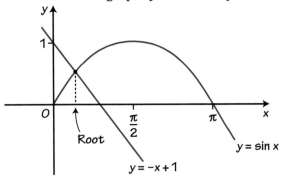

There is only one root and it is between 0 and $\frac{\pi}{2}$.

$f(x) = \sin x + x - 1$

$f'(x) = \cos x + 1$

There is only one point of intersection between the line and the curve, so this means there is only one root for the given equation.

$$x_{n+1} = x_n - \frac{f(x_n)}{f'(x_n)}$$

$x_0 = 0.5$

Try to do this calculation using the calculator with ANS. Also remember to change the calculator to radians for angles.

$$x_1 = x_0 - \frac{f(x_0)}{f'(x_0)}$$

$$= 0.5 - \frac{\sin 0.5 + 0.5 - 1}{\cos 0.5 + 1}$$

$$= 0.510957953$$

To four decimal places both x_2 and x_3 are 0.5110

$$x_2 = x_1 - \frac{f(x_1)}{f'(x_1)}$$

$$= 0.5109734294$$

$$x_3 = 0.5109734294$$

Hence root is 0.5110 correct to 4 d.p.

3 (a) (i) $\displaystyle\int_0^a (e^{2x} - 1)\,dx = \left[\frac{e^{2x}}{2} - x\right]_0^a$

$$\left[\left(\frac{e^{2a}}{2} - a\right) - \left(\frac{e^0}{2} - 0\right)\right]$$

$$= \frac{e^{2a}}{2} - a - \frac{1}{2}$$

(ii) $\dfrac{e^{2a}}{2} - a - \dfrac{1}{2} = \dfrac{1}{2}(9 - a)$

Multiply both sides by 2 to remove the 2 in the denominator.

$$e^{2a} - 2a - 1 = 9 - a$$

$$e^{2a} - a - 10 = 0$$

(b) When $a = 1$, $e^{2a} - a - 10 = e^2 - 1 - 10 = -3.61$
When $a = 2$, $e^{2a} - a - 10 = e^4 - 2 - 10 = 42.6$
As there is a sign change, this indicates a root between 1 and 2.

$$a_{n+1} = \frac{1}{2}\ln(a_n + 10)$$

$$a_1 = \frac{1}{2}\ln(1.2 + 10)$$
$$= 1.207956\ldots$$

> Use a calculator for iterations. Putting repeated values into a formula without using the iteration function of your calculator is tedious and it often results in a mistake.

$a_2 = 1.208311\ldots$
$a_3 = 1.208327\ldots$
$a_4 = 1.208328\ldots$
$a_4 = 1.20833$ correct to 5 d.p.

> We can stop at a_4 as the value correct to 5 d.p. is staying constant.

When $a = 1.208325$, $e^{2a} - a - 10 = -0.00008$
When $a = 1.208335$, $e^{2a} - a - 10 = 0.00014$
As there is a sign change, this indicates the root is 1.20833 correct to 5 d.p.

> We can see there is a zero value between these two values for a.

4 (a)

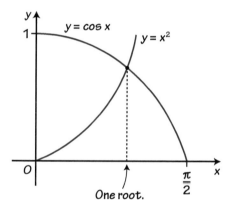

One root.

> Make sure you only draw the section of the graph between $x = 0$ and $x = \frac{\pi}{2}$.

One point of intersection in the range shows one root.

(b) At the point of intersection, $x^2 = \cos x$, so $x^2 - \cos x = 0$
Let $f(x) = x^2 - \cos x$
$f(0.8) = 0.8^2 - \cos 0.8 = -0.0567$
$f(0.9) = 0.9^2 - \cos 0.9 = 0.1883$
There is a sign change so the root a lies between these two values.

> We can equate the y-values at the point of intersection of the line and curve.

Exam practice answers

(c) $x_{n+1} = \sqrt{\cos x_n}$

As $x_0 = 0.8$

$x_1 = \sqrt{\cos x_0} = \sqrt{\cos 0.8} = 0.83468 \dots$

$x_2 = \sqrt{\cos x_1} = \sqrt{\cos 0.83468 \dots} = 0.81939 \dots$

$x_3 = \sqrt{\cos x_2} = \sqrt{\cos 0.81939 \dots} = 0.82623 \dots$

When this is continued a value to a fixed number of decimal places is obtained. This means the iteration formula can be used.

5 (a) $h = \dfrac{4-0}{n} = \dfrac{4-0}{4} = 1$

When $x = 0, y_0 = \dfrac{1}{0^2 + 2} = 0.5$

$x = 1, y_1 = \dfrac{1}{1^2 + 2} = 0.333333$

$x = 2, y_2 = \dfrac{1}{2^2 + 2} = 0.166667$

$x = 3, y_2 = \dfrac{1}{3^2 + 2} = 0.090909$

$x = 4, y_4 = \dfrac{1}{4^2 + 2} = 0.055556$

$\int_a^b y\,dx \approx \dfrac{1}{2} h\{(y_0 + y_n) + 2(y_1 + y_2 + \dots + y_{n-1})\}$

$\approx \dfrac{1}{2} h\{(y_0 + y_4) + 2(y_1 + y_2 + y_3)\}$

$\int_0^4 \dfrac{1}{x^2 + 2}\,dx \approx \dfrac{1}{2} \times 1\{(0.5 + 0.055556) + 2(0.333333$
$+ 0.166667 + 0.090909\}$

≈ 0.868667

≈ 0.9 (1 d.p.)

(b) As most of the curve lies below the trapeziums used to calculate the area the area will be smaller than that calculated in (a).

Topic 9

1 (a)

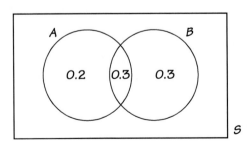

$P(A \cup B) = 0.2 + 0.3 + 0.3$
$= 0.8$

Watch out

Remember that n is the number of strips not the number of ordinates.

The formula for the Trapezium rule is obtained from the formula booklet.

206

(b) $P(A \cup B)' = 1 - P(A \cup B)$

$\qquad\qquad = 1 - 0.8$

$\qquad\qquad = 0.2$

(c) $P(B|A) = \dfrac{P(A \cap B)}{P(A)}$

$\qquad\quad = \dfrac{0.3}{0.5}$

$\qquad\quad = 0.6$

2 (a) $P(1 \text{ film of each type}) = 6 \times \dfrac{5}{10} \times \dfrac{3}{9} \times \dfrac{2}{8} = \dfrac{1}{4}$

(b) $P(3 \text{ war films}) = \dfrac{5}{10} \times \dfrac{4}{9} \times \dfrac{3}{8} = \dfrac{1}{12}$

(c) $P(3 \text{ of the same type}) = P(3 \text{ war films}) + P(3 \text{ cowboy films})$

$\qquad\qquad = \dfrac{5}{10} \times \dfrac{4}{9} \times \dfrac{3}{8} + \dfrac{3}{10} \times \dfrac{2}{9} \times \dfrac{1}{8}$

$\qquad\qquad = \dfrac{1}{12} + \dfrac{1}{120}$

$\qquad\qquad = \dfrac{11}{120}$

> Note that there are only 2 horror films so the probability of selecting 3 horror films is zero.

3 (a) $P(A') = \dfrac{66}{112} = \dfrac{33}{56}$

(b) $P(B' \cap A') = \dfrac{38}{112} = \dfrac{19}{56}$

(c)

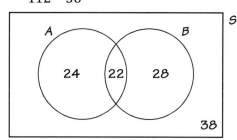

$P(A \cup B) = \dfrac{24 + 22 + 28}{112} = \dfrac{74}{112} = \dfrac{37}{56}$

> Start off completing the Venn diagram by putting in the number for the intersection of A and B (i.e. 22). To find the number for A only subtract 22 from 46 to give 24. For B we subtract 22 from 50 to give 28.
>
> We can now add 24, 22 and 28 together and subtract the total from 112 to give 38 which is placed outside the two sets.

4 (a) As events A and B are independent, then

$P(A \cup B) = P(A) \times P(B) = 0.7 \times 0.4 = 0.28$

Now $P(A \cup B) = P(A) + P(B) - P(A \cap B)$

$\qquad\qquad = 0.7 + 0.4 - 0.28$

$\qquad\qquad = 0.82$

(b) (i) Probability $= P(A) \times P(B') + P(A') \times P(B)$

$\qquad\qquad\qquad = 0.7 \times 0.6 + 0.3 \times 0.4$

$\qquad\qquad\qquad = 0.54$

(ii) $P(A' \text{ and } B') = P(A') \times P(B') = 0.3 \times 0.6 = 0.18$

> Note that we use this formula because we are told the events are independent.

5   (a) $P(A \cup B) = P(A) + P(B) - P(A \cap B)$
           $P(A \cap B) = P(A) + P(B) - P(A \cup B)$
                      $= 0.7 + 0.5 - 0.85$
                      $= 0.35$
        Now if the events are dependent, $P(A \cap B) = P(A) \times P(B)$
                                       $= 0.7 \times 0.5$
                                       $= 0.35$
        As $P(A) \times P(B) = P(A \cap B)$, A and B are independent events.
   (b) $P(A$ only$) = P(A) - P(A \cap B) = 0.7 - 0.35 = 0.35$
        $P(B$ only$) = P(B) - P(A \cap B) = 0.5 - 0.35 = 0.15$
        $P(A$ or B only$) = 0.35 + 0.15 = 0.5$

6   (a) $P(M) = \dfrac{52}{100} = \dfrac{13}{25}$

   (b) $P(M \cap G) = \dfrac{28}{100} = \dfrac{7}{25}$ $P(M \cap G)$ is the probability that a student selected at random takes maths and is a girl.

   (c) $P(M \mid G) = \dfrac{28}{54} = \dfrac{14}{27}$ $P(M \mid G)$ is the probability that a student selected at random takes maths given that they are a girl. Notice that we are now only picking from the 54 girls.

7   (a)   P (two people have virus) $= 0.06 \times 0.06 = 0.0036$
   (b)   Let V = the event the person has the virus.
        Let P = the event that a positive response is obtained.

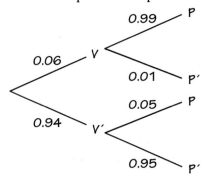

        $P(V' \cap P') = 0.94 \times 0.95 = 0.893$

   (c)   P(positive result) = P(they have the virus and a positive) +
        P(they don't have the virus and a positive)
                            $= 0.06 \times 0.99 + 0.94 \times 0.05$
                            $= 0.1064$

(d) We need the probability of V given that P has occurred.

$$P(V|P) = \frac{P(V \cap P)}{P(P)}$$

$$= \frac{0.06 \times 0.99}{0.1064}$$

$$= 0.558$$

> This is conditional probability so the formula
> $$P(B|A) = \frac{P(A \cap B)}{P(A)}$$
> obtained from the formula booklet is adapted to give
> $$P(V|P) = \frac{P(V \cap P)}{P(P)}.$$

Topic 10

1 (a) The distribution consists of discrete data and each number has an equal probability of occurring. Hence use a discrete uniform probability distribution.

(b) As distances are continuous data and the data is likely to be symmetrical about a mean value the normal distribution can be used.

(c) Times are continuous data, and the data is likely to be symmetrical about a mean value the normal distribution can be used.

(d) Salaries are continuous data and are likely to be symmetrical about a mean value so a normal distribution can be used.

(e) The number of cuttings growing is discrete data and the probability of a cutting successfully growing is fixed and independent. The binomial distribution can be used.

(f) The data is continuous and symmetrical about the mean and it is bell-shaped. A normal distribution can be used.

2 (a) Mean $E(X) = \frac{1}{2}(a + b) = \frac{1}{2}(3 + 6) = 4.5$ mins

(b) Variance $\text{Var}(X) = \frac{1}{12}(b - a)^2 = \frac{1}{12}(6 - 3)^2 = 0.75$ mins

(c) $P(3 \leq X \leq 5) = \frac{1}{6 - 3} \times 2 = 0.67$

3 (a)

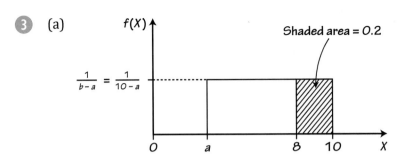

$P(X \geq 8)$ = shaded area

$$= \frac{1}{10-a} \times 2$$

Now $P(X \geq 8) = 0.2$

$$= \frac{1}{10-a} \times 2 = 0.2$$
$$2 = 0.2(10-a)$$
$$2 = 2 - 0.2a$$
$$0 = 0.2a$$
$$a = 0$$

(b) Variance Var $(X) = \frac{1}{12}\left(b-a\right)^2 = \frac{1}{12}\left(10-0\right)^2 = 8.33$

(c) $P(X \leq 5) = \frac{1}{10} \times 5 = 0.5$

(d) $P(2 \leq X \leq 7) = \frac{1}{10} \times 5 = 0.5$

4 If X is the length of the string to the cut measured from the left to right, then the other length will be $10 - X$.
Now if X is over 7 cm then this will be what is required.
However, if X is less than 3 cm then the other piece now becomes the longest piece over 7cm, so this is also what is required.
Hence, we require $P(X > 7)$ or $P(X < 3)$.

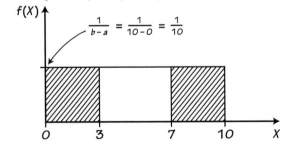

Probability that length of longest piece > 7 cm
  = sum of shaded areas
$$= \frac{1}{10} \times 3 + \frac{1}{10} \times 3$$
$$= 0.6$$

5 (a)

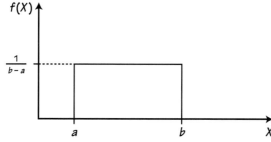

Mean $E(X) = \frac{1}{2}(a + b) = 5$, hence $a + b = 10$ (1)

Variance $\text{Var}(X) = \frac{1}{12}(b - a)^2 = 3$

$$(b - a)^2 = 36$$

Square rooting both sides, we obtain $b - a = \pm 6$

Now $b > a$ so $b - a = 6$ (2)

Adding equations (1) and (2), we obtain:

$2b = 16, b = 8$

$a + b = 10$

$a + 8 = 10$

$\quad a = 2$

> Notice that we do not know either a or b, so we need to find two equations and then solve them simultaneously.

(b)

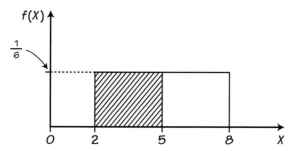

$P(X \le 5) =$ shaded area

$\qquad = \frac{1}{6} \times 3$

$\qquad = \frac{1}{2}$

6 (a)

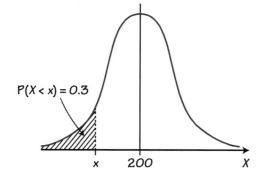

$P(X < x) = 0.3$

> As the probability is 0.3, the value of x will be in the lower tail as this tail covers probabilities from infinitely small to 0.5

Use Inverse Normal on the calculator and enter the following parameters:
Area: 0.3
σ: 4
μ: 200
The calculator gives a value xInv =197.9023 ...
The value of x is 197.90

> Remember that we need to work out the probability from the lower tail to find the value of x.

(b)

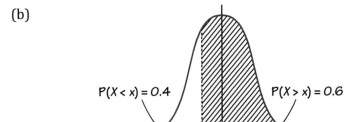

Use Inverse Normal on the calculator and enter the following parameters:
Area: 0.4
σ: 4
μ: 200
The calculator gives a value xInv =198.9866 ...
The value of x is 198.99.

7 (a) We need to find $P(X > 50)$ when $\mu = 34$ and $\sigma = 11$.

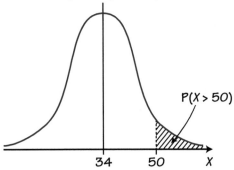

To find $P(X > 50)$ using the CASIO Classwiz calculator, take the following steps:
Select 'Distribution' and then 'Normal CD'
Now enter the following parameters:
Lower: 50
Upper: 1×10^{99}
σ: 11
μ: 34
The answer is displayed as P = 0.07289 ...

(b)

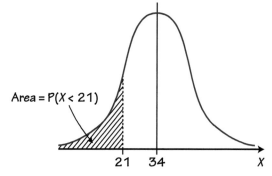

To find P(X < 21) using the CASIO Classwiz calculator, take
the following steps:
Select 'Distribution' and then 'Normal CD'
Now enter the following parameters:
Lower: 1×10^{-99}
Upper: 21
σ: 11
μ: 34
The answer is displayed as P = 0.11864 ...

(c)

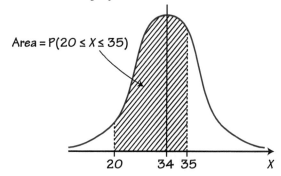

To find P($20 \le X \le 35$) using the CASIO Classwiz calculator,
take the following steps:
Select 'Distribution' and then 'Normal CD'
Now enter the following parameters:
Lower: 20
Upper: 35
σ: 11
μ: 34
The answer is displayed as P = 0.4346 ...

⑧ (a) Normal distribution $X \sim N(685, 5^2)$

(b)

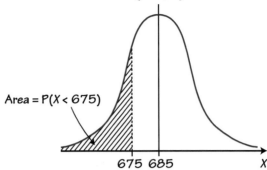

Area = P(X < 675)

675 685   X

To find P(X < 675) using the CASIO Classwiz calculator, take the following steps:

Select 'Distribution' and then 'Normal CD'

Now enter the following parameters:

Lower: 1×10^{-99}

Upper: 675

σ: 5

μ: 685

The answer is displayed as P = 0.0227 ...

(c)

> The Standard Normal Distribution needs to be used for this part of the question.

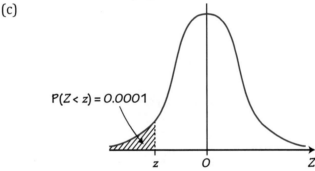

P(Z < z) = 0.0001

z   O   Z

Use the 'Inverse Normal' on the calculator with $\mu = 0, \sigma = 1$ and P(Z < z) = area shaded = 0.0001.

The calculator gives a z-value of −3.7190.

> Both these formulas are included in the formula booklet.

When x = 675 and σ = 5 $z = \dfrac{x - \mu}{\sigma}$

Hence $-3.7190 = \dfrac{675 - \mu}{5}$

$\mu = 693.60\,g$

⑨ (a) (i) A continuous uniform distribution with parameters [2, 10] as intervals of equal time are equally likely to occur between 2 and 10 minutes.

(ii) Mean $E(X) = \dfrac{1}{2}\left(a + b\right) = \dfrac{1}{2}\left(2 + 10\right) = 6$ mins

Variance Var $(X) = \dfrac{1}{12}\left(b - a\right)^2 = \dfrac{1}{12}\left(10 - 2\right)^2 = 5.33$ mins

(iii) Any time between 2 and 10 minutes is equally likely to occur.

(b) (i) P(Nicky takes the call) = $\dfrac{1}{5}$

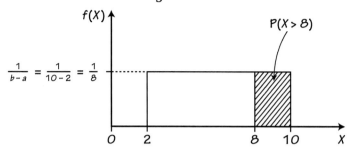

$$P(X > 8) = \dfrac{1}{8} \times 2 = \dfrac{1}{4}$$

P(Nicky takes call and call is over 8 mins) = $\dfrac{1}{5} \times \dfrac{1}{4} = \dfrac{1}{20}$

> The AND rule is used here so we multiply the probabilities.

(ii) P(Nicky does not take the call) = $1 - \dfrac{1}{5} = \dfrac{4}{5}$

$$P(5 < X < 7) = \dfrac{1}{8} \times 2 = \dfrac{1}{4}$$

P(Nicky does not take the call and call takes between 5 and 7 mins) = $\dfrac{4}{5} \times \dfrac{1}{4} = \dfrac{1}{5}$

10 (a) The masses are continuous data, and the values are grouped symmetrically around the mean value. The distribution is tallest in the middle and tails off equally either side towards the ends.

(b) (i) To find $P(X > 75)$ using the CASIO Classwiz calculator, take the following steps:
Select 'Distribution' and then 'Normal CD'
Now enter the following parameters:
Lower: 75
Upper: 1×10^{99}
σ: 10
μ: 47.5
The answer is displayed as $P = 2.9797 \ldots \times 10^{-3}$
Hence probability = 0.003
Number of potatoes having mass over 75 g
= 0.003 × 100 = 0.3

(ii) To find P(35≤X ≤ 45) using the CASIO Classwiz calculator, take the following steps:
Select 'Distribution' and then 'Normal CD'
Now enter the following parameters:
Lower: 35
Upper: 45
σ: 10
μ: 47.5
The answer is displayed as P = 0.2956...
Hence probability = 0.296
Number of potatoes with masses in range $35 \le m \le 45$
= 0.296 × 100 = 29.6

(c) For $m > 75$ the model predicts 0.3 potatoes and the table gives 0, so this is pretty close.
For $35 \le m \le 45$ the model predicts 29.6 potatoes and the table gives 29 so they approximately agree.
Hence from these conclusions, the model could be suitable.

(d) Next year might be different as the weather and growing conditions could be different (e.g. drought, too much rain, frost, disease, insects, etc.), so the model would be unlikely to give an accurate prediction.

11 (a) The marks are random continuous data, and the distribution is symmetrical about the mean mark.

(b)

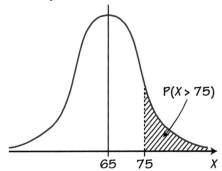

Use Normal CD
Lower: 75
Upper: 1×10^{99}
σ: 6
μ: 65
Probability = 0.04779

(c)

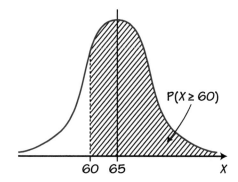

Use Normal CD
Lower: 60
Upper: 1×10^{99}
σ: 6
μ: 65
Probability = 0.7977
Number of students passing = 0.7977×70
$\qquad\qquad\qquad\qquad\qquad$ = 55.8
$\qquad\qquad\qquad\qquad\qquad$ = 56 (nearest integer)

(d)

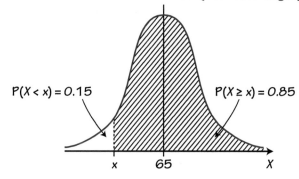

Use Inverse Normal
Area: 0.15
σ: 6
μ: 65
xInv = 58.7814
Pass mark needs to be 59 marks (nearest integer)

Topic 11

1. There are two possible ways to perform a hypothesis test; using a critical value or using a p-value. Here we will perform the calculation using both methods.
   Solution using a critical value
   (a) $H_0 : \mu = 20$ $\qquad$ $H_1 : \mu > 20$

(b) Assuming H_0, if $X \sim N(\mu, \sigma^2)$ then the sample mean weight
$$\overline{X} \sim N\left(20, \frac{0.769^2}{10}\right)$$

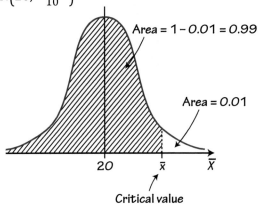

Area = 1 − 0.01 = 0.99

Area = 0.01

20     $\overline{x}$ $\overline{X}$

Critical value

Use Inverse Normal with the following parameters:
Area: 0.99 (i.e. 1 − 0.01 = 0.99)
$$\sigma: \frac{0.769}{\sqrt{10}} = 0.2432$$
μ: 20
Gives a critical value of 20.5658
Now, sample mean 20.6 > critical value 20.5658 so it
lies in the critical region.
This means that there is sufficient evidence to
conclude that the null hypothesis should be rejected.
This means there is evidence that the addition of the
chemical does in fact improve the breaking strength of
the cable.

Here is the same question answered using p-values
(a) $H_0 : \mu = 20$ $H_1 : \mu > 20$
(b) Using a calculator to work out the p-value.

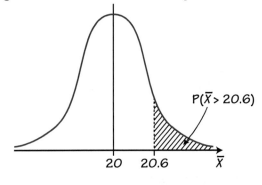

$P(\overline{X} > 20.6)$

20   20.6 $\overline{X}$

The sample mean
(i.e. 20.6) is the test
statistic. We then look
to see if it is greater
than the critical value
which puts in in the
critical region.

Set the calculator to 'Normal CD' and type in:
Lower = 20.6
Upper = 1×10^{99}
$\sigma = \dfrac{0.769}{\sqrt{10}} = 0.2432$
$\mu = 20$
p-value = $P(\overline{X} > 20.6) = 0.0068$
As the p-value < 0.01 (i.e. the significance level) there is sufficient evidence to conclude that the null hypothesis should be rejected. This means there is evidence that the addition of the chemical does in fact improve the breaking strength of the cable.

2 (a) Null hypothesis is $H_0 : \rho = 0$
Alternative hypothesis is $H_1 : \rho < 0$

(b) This is a one-tailed test, so looking up 5% significance level for a one-tailed test and $n = 15$, the critical value is read off from the 'Critical values of the product moment correlation coefficient table'.

For the null hypothesis we say there is no correlation, so the correlation coefficient is zero.

TABLE 9 CRITICAL VALUES OF THE PRODUCT MOMENT CORRELATION COEFFICIENT

The table gives the critical values, for different significance levels, of the sample product moment correlation coefficient r based on n independent pairs of observations from a bivariate normal distribution with correlation coefficient $\rho = 0$.

One tail	10%	5%	2.5%	1%	0.5%
Two tail	20%	10%	5%	2%	1%
n					
4	0.8000	0.9000	0.9500	0.9800	0.9900
5	0.6870	0.8054	0.8783	0.9343	0.9587
6	0.6084	0.7293	0.8114	0.8822	0.9172
7	0.5509	0.6694	0.7545	0.8329	0.8745
8	0.5067	0.6215	0.7067	0.7887	0.8343
9	0.4716	0.5822	0.6664	0.7498	0.7977
10	0.4428	0.5494	0.6319	0.7155	0.7646
11	0.4187	0.5214	0.6021	0.6851	0.7348
12	0.3981	0.4973	0.5760	0.6581	0.7079
13	0.3802	0.4762	0.5529	0.6339	0.6835
14	0.3646	0.4575	0.5324	0.6120	0.6614
15	0.3507	0.4409	0.5140	0.5923	0.6411
16	0.3383	0.4259	0.4973	0.5742	0.6226

The alternative hypothesis is that there is negative correlation, so the correlation coefficient is less than zero.

The critical value is shown highlighted in the above table. Notice that the positive value 0.4409 is given in the table.
Critical value = 0.4409
Now the PMCC = −0.87 but for comparison we change it to a positive value and as 0.87 > 0.4409 we can say the result is significant.
Hence at the 5% level of significance there is evidence to reject the null hypothesis in favour of the alternative hypothesis that the age of a chicken and the number of eggs they produce are negatively correlated.

Remember to say there is evidence to reject/accept the null hypothesis. Never say that the null hypothesis or alternative hypothesis should be simply 'accepted'.

The sample mean is used as the test statistic to see if it lies in the critical region.

3 (a) $H_0 : \mu = 4.5$ $\quad$ $H_1 : \mu \neq 4.5$

(b) Let X be the mean weight of a bird which is normally distributed.

Assuming H_0, if $X \sim N(\mu, \sigma^2)$ then the sample mean weight $\overline{X} \sim N\left(\mu, \frac{\sigma^2}{n}\right)$

Sample mean = 4.2

Using critical values

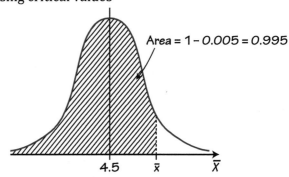

Area = 1 − 0.005 = 0.995

4.5 $\quad$ $\overline{x}$ $\quad$ $\overline{X}$

Remember that the area is measured from the lower tail.

Use Inverse Normal with the following parameters:

Area: 0.995 (i.e. 1 − 0.005 = 0.995)

$\sigma: \dfrac{0.1656}{\sqrt{10}} = 0.0524$

μ: 4.5

Gives a critical value, $\overline{X}$ of 4.6350 for the upper tail.

Since the distribution is symmetrical this gives a value of 4.6350 − 4.5 = 0.1350 from the mean so this gives a critical value of 4.5 − 0.1350 = 4.365 for the lower tail.

The critical regions are $\overline{X} < 4.365$ and $\overline{X} > 4.6350$.

The test statistic, the sample mean = 4.2 kg so it lies in the critical region.

Hence there is evidence to reject the null hypothesis and conclude that the mean weight of the birds is not 4.5 kg.

4 (a) The volume of beer is a continuous variable which produced a bell-shaped distribution which is symmetrical about the mean.

(b) $H_0 : \mu = 445$ $\quad$ $H_1 : \mu < 445$

(c) Let X be the volume of beer in a can, which is normally distributed.

Assuming H_0, if $X \sim N(\mu, \sigma^2)$ so $X \sim N(445, 8^2)$

The test statistic is the sample mean, $\overline{X}$.

Now $\bar{X} \sim N\left(445, \frac{8^2}{40}\right)$ so $\bar{X} \sim N\left(445, 1.6\right)$

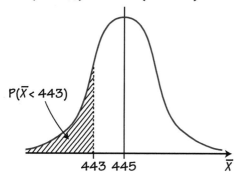

$P(\bar{X} < 443)$

443 445 $\bar{X}$

Use Normal CD
Lower = 1×10^{-99}
Upper = 443
$\sigma = 8 \div \sqrt{40} = 1.2649$
$\mu = 445$
p-value = 0.05692
Now $0.05692 > 0.05$
There is insufficient evidence to reject the null
hypothesis H_0 and we conclude that the customers
complaints are unjustified

The probability of obtaining a value of less than 443 (i.e. 0.1587) is greater than the significance level (i.e. 0.05) so we do not reject the null hypothesis.

5. (a) There is weak negative correlation between intelligence
   and daily hours of watching TV.
   (b) Null hypothesis $H_0:r = 0$ (i.e. no correlation)
   Alternative hypothesis $H_1:r < 0$ (i.e. negative correlation)
   We are performing a one-tailed test and the sample
   size, $n = 10$. The significance level is 1%.
   We then use the tables 'Critical values of the product
   moment correlation coefficient'.
   Look up $n = 10$ and significance level = 1% for a one-
   tailed test.
   Reading off we obtain a critical value = 0.7155
   Now we know we are investigating negative correlation,
   so critical value = −0.7155
   Now the PMCC = −0.36 and as −0.36 > −0.7155 we can
   say the result is not significant.
   Hence, we fail to reject the null hypothesis owing
   to there being insignificant evidence. This means
   there is insufficient evidence to conclude that higher
   intelligence is associated with lower number of daily
   hours of TV viewing so her conclusion is correct.

This question has been answered using *p*-values

6 (a) We can draw a sketch of what we know:

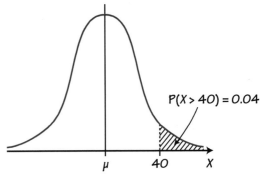

$$P(X > 40) = 0.04$$

As we do not know the mean, we need to use the standard normal distribution and find the *z*-value.

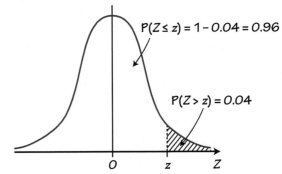

$$P(Z \leq z) = 1 - 0.04 = 0.96$$

$$P(Z > z) = 0.04$$

We need to use 0.96 for the area as the area is measured from the lower tail toward the upper tail.

Using the calculator:
Choose Inverse Normal
Enter the following parameters:
Area = 0.96
σ: 1
μ: 0
Result gives *z*-value = 1.7507

We now use the formula $z = \dfrac{x - \mu}{\sigma}$

$$1.7507 = \frac{40 - \mu}{5}$$
$$1.7507 \times 5 = 40 - \mu$$
$$\mu = 31.25 \text{ mins}$$

(b)

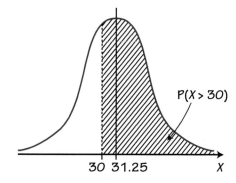

$P(X > 30)$

30 31.25 X

Use Normal CD
Enter the following parameters:
Lower: 30
Upper: = 1×10^{99}
σ: 5
μ: 31.25
Probability (i.e. p-value) = 0.5987

> We are now finding $P(X>30)$ by finding the area under the curve between 30 and a very large value (i.e. 1×10^{99})

(c) $H_0 : \mu = 31.25$ $H_1 : \mu < 31.25$
Let $\bar{X}$ be the mean time per treatment which is normally distributed.
Now $\bar{X} \sim N\left(31.25, \frac{5^2}{20}\right)$
We will now use a calculator to work out the p-value.
Use Normal CD
Enter the following parameters:
Lower: = 1×10^{-99}
Upper: 30
σ: $\dfrac{5}{\sqrt{20}} = 1.118$
μ: 31.25
Probability (i.e. p-value) = 0.1318
p-value 0.1318 > 0.05 (i.e. the significance level) there is evidence at the 5% level of significance to fail to reject the null hypothesis. Hence, there is no evidence to conclude that the new dental equipment decreases the treatment time.

This question has been answered using critical values

   (a) (i) $H_0 : \mu = 9$ $H_1 : \mu \neq 9$

Exam practice answers

This states the diameter of the population is normally distributed with mean 9 and variance 0.35^2.

This states the sample mean diameters are normally distributed with mean 9 and variance $\frac{0.35^2}{20}$.

Remember that we are performing a two-tailed test, so we halve the significance level (so 0.005 at both ends).

(ii) Assuming $H_0, D \sim N(9, 0.35^2)$

Now $\bar{D} \sim N\left(9, \frac{0.35^2}{20}\right)$

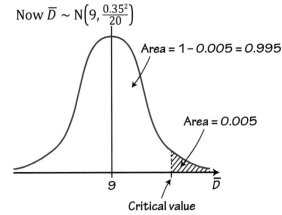

Area = 1 – 0.005 = 0.995

Area = 0.005

9

Critical value

Use Inverse Normal with the following parameters:
Area: 0.995 (i.e. 1 – 0.005 = 0.995)
σ: $\frac{0.35}{\sqrt{20}} = 0.07826$
μ: 9

This gives a critical value of 9.2016. The critical region is symmetrical on either side of the mean, so the critical region is < 8.7984 and > 9.2016.

(b) Now the sample mean (i.e. 9.1) is the test statistic. This test statistic is not in the critical region, so the result is not significant, and we fail to reject the null hypothesis and conclude that there is no evidence that the machine is malfunctioning.

Topic 12

1 $v = 9t^2 - 6t + 1$

$\frac{dv}{dt} = 18t - 6$

Minimum velocity occurs when $\frac{dv}{dt} = 0$.

$18t - 6 = 0$

$t = \frac{1}{3}$ s

Hence time when moving with minimum velocity $= \frac{1}{3}$ s

$s = \int v \, dt$

$\quad = \int (9t^2 - 6t + 1) \, dt$

$\quad = \frac{9t^3}{3} - \frac{6t^2}{2} + t + c$

$\quad = 3t^3 - 3t^2 + t + c$

Note that as there is only one time, we do not have to test whether it is a maximum or a minimum.

When $t = 0$, $s = 0$, so
$0 = 3(0)^3 - 3(0)^2 + (0) + c$
Giving $c = 0$
Hence $s = 3t^3 - 3t^2 + t$
When $t = \dfrac{1}{3}$ s, $s = 3\left(\dfrac{1}{3}\right)^3 - 3\left(\dfrac{1}{3}\right)^2 + \dfrac{1}{3}$

$$= \dfrac{1}{9} \text{ m}$$

2 (a) $v = 2t(2 - 3t + t^2)$
$\quad = 4t - 6t^2 + 2t^3$
$a = \dfrac{\mathrm{d}v}{\mathrm{d}t}$
$\quad = 4 - 12t + 6t^2$
$\quad = 6t^2 - 12t + 4$

(b) $s = \int v\,\mathrm{d}t$

$\quad = \int 4t - 6t^2 + 2t^3\,\mathrm{d}t$

$\quad = 2t^2 - 2t^3 + \dfrac{t^4}{2} + c$

When $t = 0$, $s = 3$ so $c = 3$

$s = 2t^2 - 2t^3 + \dfrac{t^4}{2} + 3$

When $t = 2$, $s = 2(2)^2 - 2(2)^3 + \dfrac{(2)^4}{2} + 3$

$\quad = 8 - 16 + 8 + 3$

$\quad = 3 \text{ m}$

3 (a) $v = 2t^2 - 18t + 36$
$\quad 0 = t^2 - 9t + 18$
$\quad = (t - 3)(t - 6)$
Hence particle is at instantaneous rest at $t = 3$ s and $t = 6$ s.

(b) $\dfrac{\mathrm{d}v}{\mathrm{d}t} = 4t - 18$

At the minimum value of v, $\dfrac{\mathrm{d}v}{\mathrm{d}t} = 0$.

$4t - 18 = 0$ so $t = \dfrac{9}{2}$ s

$v = 2t^2 - 18t + 36$

$\quad = 2\left(\dfrac{9}{2}\right)^2 - 18\left(\dfrac{9}{2}\right) + 36$

$\quad = -\dfrac{9}{2} \text{ ms}^{-1}$

Exam practice answers

(c)

Note that the area between the limits 6 and 3 is negative as the area is below the x-axis. As we require the distance, we change this to a positive area.

$$\text{Distance} = \int_2^3 (2t^2 - 18t + 36)\,dt + \int_3^6 (2t^2 - 18t + 36)\,dt$$

$$= \left[\frac{2t^3}{3} - 9t^2 + 36t\right]_2^3 + \left[\left(\frac{2t^3}{3} - 9t^2 + 36t\right)\right]_3^6$$

$$= \left[(45) - \left(\frac{124}{3}\right)\right] + [(36) - (45)]$$

$$= \frac{11}{3} - 9 \quad \text{(remember to change the negative area to a positive one)}$$

$$= \frac{11}{3} + 9$$

$$= 12.67\text{m}$$

4. (a) $a = 4 - 6t$

$$v = \int a\,dt$$

$$= \int (4 - 6t)\,dt$$

$$= 4t - \frac{6t^2}{2} + c$$

$$= 4t - 3t^2 + c$$

When $t = 0$, $v = 4$

$4 = 4(0) - 3(0) + c$

$c = 4$

Hence, $v = 4t - 3t^2 + 4$

(b) $s = \int v\,dt$

$$= \int (4t - 3t^2 + 4)\,dt$$

$$= \frac{4t^2}{2} - \frac{3t^3}{3} + 4t + c$$

$$= 2t^2 - t^3 + 4t + c$$

When $t = 0$, $s = 0$

$0 = 2(0)^2 - (0)^3 + 4(0) + c$

$c = 0$

Hence, $s = 2t^2 - t^3 + 4t$

(c) When particle is at rest, $v = 0$.

$$0 = 4t - 3t^2 + 4$$
$$3t^2 - 4t - 4 = 0$$
$$(3t + 2)(t - 2) = 0$$
$$t = 2 \text{ s}$$

When $t = 2$ s, $s = 2(2)^2 - (2)^3 + 4(2) = 8$ m

(d) When $t = 3$ s, $v = 4t - 3t^2 + 4$
$$= 4(3) - 3(3)^2 + 4$$
$$= -11 \text{ ms}^{-1}$$

Speed $= 11$ ms^{-1}
Acceleration, $a = 4 - 6t$
When $t = 3$s, $a = 4 - 6(3) = -14$ ms^{-2}
Now as the particle velocity is negative it is moving in the opposite direction and as the acceleration is also negative, it is acting in the same direction. This means the particle speed is increasing.

⑤ (a) $s = \int v \, dt$

$$= \int (8t - 18t^2) \, dt$$
$$= \frac{8t^2}{2} - \frac{18t^3}{3} + c$$
$$= 4t^2 - 6t^3 + c$$

When $t = 1$, $s = 0$
$$0 = 4(1)^2 - 6(1)^3 + c$$
$$c = 2$$
Hence, $s = 4t^2 - 6t^3 + 2$

(b) $a = \dfrac{dv}{dt}$
$$= 8 - 36t$$

The solution $t = -\frac{2}{3}$ s is ignored as t cannot be negative.

Note that the negative velocity means the particle is travelling in the opposite direction.

Speed is a scalar quantity, so we do not include the minus sign. We are only concerned with the magnitude and not the direction.

6 (a) To obtain the graph we can look at the transformations that need to be made to the graph of $y = \sin t$ to give the graph of $y = -3 \sin \dfrac{t}{2}$.

The minus means a reflection in the x-axis.

The 3 means the curve will go down as far as -3.

The $\dfrac{t}{2}$ means a stretch parallel to the x-axis with scale factor 2.

When sketching the graph, make sure you only include the section of the graph for the range $0 \le t \le \pi$.

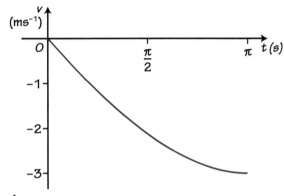

(b) $s = \int v \, dt$

$\quad = \displaystyle\int_0^{\frac{\pi}{2}} \left(-3 \sin \frac{t}{2}\right) dt$

$\quad = \left[6 \cos \dfrac{t}{2}\right]_0^{\frac{\pi}{2}}$

$\quad = \left[\left(6 \cos \dfrac{\pi}{4}\right) - (6 \cos 0)\right]$

$\quad = \dfrac{6}{\sqrt{2}} - 6$

$\quad = -1.76 \, \text{m}$

Distance $= 1.76 \, \text{m}$

We have calculated the displacement but note the question asks for the distance, so the negative sign is removed.

Topic 13

1 (a) $\mathbf{a} = \dfrac{\mathbf{F}}{m}$

$\quad = \dfrac{1}{3}(6\mathbf{i} + 9\mathbf{j})$

$\quad = 2\mathbf{i} + 3\mathbf{j}$

$\mathbf{v} = \mathbf{u} + \mathbf{a}t$

$\quad = (\mathbf{i} + 4\mathbf{j}) + (2\mathbf{i} + 3\mathbf{j})(2)$

$\quad = 5\mathbf{i} + 10\mathbf{j}$

You are given the force in the question so that the acceleration can be found. The value for the acceleration can then be used in the equations of motion.

The force is constant so the acceleration will be constant.

(b) $\mathbf{s} = \mathbf{u}t + \frac{1}{2}\mathbf{a}t^2$

$\quad = (\mathbf{i} + 4\mathbf{j})(2) + \frac{1}{2}(2\mathbf{i} + 3\mathbf{j})(2^2)$

$\quad = 2\mathbf{i} + 8\mathbf{j} + 4\mathbf{i} + 6\mathbf{j}$

$\quad = 6\mathbf{i} + 14\mathbf{j}$

Displacement from origin $O = \mathbf{i} - 3\mathbf{j} + 6\mathbf{i} + 14\mathbf{j}$

$\quad\quad\quad\quad\quad\quad\quad = 7\mathbf{i} + 11\mathbf{j}$

Distance from $O = \sqrt{7^2 + 11^2}$

$\quad\quad\quad\quad\quad = 13 \text{ m} \quad (2 \text{ s.f.})$

> It would be easy to give the displacement as your answer but the question asks for the distance. Always check with the question.

2 (a) $\mathbf{v} = \dfrac{d\mathbf{r}}{dt} = 4t\mathbf{i} + (2t - 2)\mathbf{j}$

When $t = 4$, $\mathbf{v} = 16\mathbf{i} - 6\mathbf{j}$

Speed $= |\mathbf{v}| = \sqrt{16^2 + (-6)^2} = 17.1 \text{ ms}^{-1}$

(b) $\mathbf{a} = \dfrac{d\mathbf{v}}{dt} = 4\mathbf{i} + 2\mathbf{j}$

As the acceleration is independent of t it means the acceleration is constant.

Magnitude of the acceleration $= |\mathbf{a}| = \sqrt{4^2 + (2)^2}$

$\quad\quad\quad\quad\quad\quad\quad\quad\quad\quad = \sqrt{20} = 4.5 \text{ ms}^{-2}$

3 (a) $OP^2 = (2t - 5)^2 + (t - 3)^2 + (7 - 2t)^2$

$\quad\quad\quad = 4t^2 - 20t + 25 + t^2 - 6t + 9 + 49 - 28t + 4t^2$

$\quad\quad\quad = 9t^2 - 54t + 83$

$\dfrac{dOP^2}{dt} = 18t - 54$

At the minimum distance for OP, $\dfrac{dOP^2}{dt} = 0$

Hence, $18t - 54 = 0$

$\quad\quad\quad\quad t = 3\text{s}$

> We need to find the value of t that will give a minimum value for OP^2.

(b) $\mathbf{r} = (2t - 5)\mathbf{i} + (t - 3)\mathbf{j} + (7 - 2t)\mathbf{k}$

$\mathbf{v} = \dfrac{d\mathbf{r}}{dt}$

$\quad = 2\mathbf{i} + \mathbf{j} - 2\mathbf{k}$

$|\mathbf{v}| = \sqrt{2^2 + 1^2 + (-2)^2}$

$\quad = \sqrt{4 + 1 + 4}$

$\quad = 3 \text{ ms}^{-1}$

> Only the positive value of $\sqrt{9}$ is used as we are only interested in the magnitude of the velocity and not its direction.

4 (a) $\mathbf{a} = \dfrac{d\mathbf{v}}{dt} = -4\sin 2t\,\mathbf{i} - \cos t\,\mathbf{j}$

(b) $\mathbf{s} = \int \mathbf{v}\,dt$

$\quad = \int (2\cos 2t\,\mathbf{i} - \sin t\,\mathbf{j})\,dt$

$\quad = \dfrac{2\sin 2t\,\mathbf{i}}{2} + \cos t\,\mathbf{j} + c$

$\quad = \sin 2t\,\mathbf{i} + \cos t\,\mathbf{j} + c$

When $t = 0$, $\mathbf{s} = 2\mathbf{i} - \mathbf{j}$

$2\mathbf{i} - \mathbf{j} = \sin 0\,\mathbf{i} + \cos 0\,\mathbf{j} + c$

$2\mathbf{i} - \mathbf{j} = 0\mathbf{i} + \mathbf{j} + c$

$\quad\quad c = 2\mathbf{i} - 2\mathbf{j}$

Hence $s = \sin 2t\,\mathbf{i} + \cos t\,\mathbf{j} + 2\mathbf{i} - 2\mathbf{j}$
$$= (2 + \sin 2t)\mathbf{i} + (\cos t - 2)\mathbf{j}$$

When $t = \dfrac{\pi}{6}$

$s = (2 + \sin 2t)\mathbf{i} + (\cos t - 2)\mathbf{j}$

$$= \left(2 + \sin\frac{2\pi}{6}\right)\mathbf{i} + \left(\cos\frac{\pi}{6} - 2\right)\mathbf{j}$$

$$= \left(2 + \sin\frac{\pi}{3}\right)\mathbf{i} + \left(\cos\frac{\pi}{6} - 2\right)\mathbf{j}$$

$$= \left(2 + \frac{\sqrt{3}}{2}\right)\mathbf{i} + \left(\frac{\sqrt{3}}{2} - 2\right)\mathbf{j}$$

5 (a) $\mathbf{r}_A = (\mathbf{i} - 10\mathbf{k}) + t(-2\mathbf{i} - 2\mathbf{j} - 5\mathbf{k})$
$$= (1 - 2t)\mathbf{i} + (-2t)\mathbf{j} + (-10 - 5t)\mathbf{k}$$
$\mathbf{r}_B = (7\mathbf{i} + 9\mathbf{j} - 6\mathbf{k}) + t(\mathbf{i} - 8\mathbf{j} - 5\mathbf{k})$
$$= (7 + t)\mathbf{i} + (9 - 8t)\mathbf{j} + (-6 - 5t)\mathbf{k}$$

(b) When $t = 2$, $\mathbf{r}_A - \mathbf{r}_B = (-3 - 9)\mathbf{i} + (-4 + 7)\mathbf{j} + (-20 + 16)\mathbf{k}$
$$= -12\mathbf{i} + 3\mathbf{j} - 4\mathbf{k}$$

Distance between A and B $= \sqrt{(-12)^2 + (3)^2 + (-4)^2}$
$$= \sqrt{144 + 9 + 16}$$
$$= \sqrt{169}$$
$$= 13 \text{ m}$$

The position vector
is the sum of the
position vector of the
starting point and the
displacement vector.
The displacement
vector can be found
by multiplying the
velocity vector by
time t.

Topic 14

1 (a) Taking moments about B, we obtain:
$$8g \times 0.3 = 0.5Q$$
$$Q = 4.8g$$
$$= 4.8 \times 9.8$$
$$= 47.04$$
$$= 47 \text{ N (2 s.f.)}$$

Resolving vertically, we obtain:
$$P + Q = 8g$$
$$P + 4.8g = 8g$$
$$P = 3.2g$$
$$= 3.2 \times 9.8$$
$$= 31.36$$
$$= 31 \text{ N (2 s.f.)}$$

(b) Let the magnitude of the weight placed at D $= W$ N.
Taking moments about C, we obtain:
$$W \times 0.3 + P \times 0.5 = 8g \times 0.2$$
$$0.3W + 0.5P = 1.6g$$
$$P = 3.2g - 0.6W$$

Another method would
be to take moments
about C:
$$0.5P = 0.2 \times 8g$$
$$P = 31.36$$
$$P = 31\,\text{N (2 s.f.)}$$

Note P will change
value from the value
found in part (a).

Now, for the plank not to lift, the normal reaction at P must be greater than or equal to zero.

Hence, $3.2g - 0.6W \geq 0$

Greatest value of $W = \dfrac{3.2g}{0.6}$

$\qquad\qquad\qquad = 52.26$

$\qquad\qquad\qquad = 52$ N (2 s.f.)

② Resolving forces in the horizontal direction:

$X = 3 + 16 \cos 60° - 10 \cos 60°$

$\quad = 3 + 8 - 5$

$\quad = 6$ N

We will take the direction to the right as the positive direction.

Resolving forces in the perpendicular direction:

$Y = 11\sqrt{3} + 10 \sin 60° + 16 \sin 60°$

$\quad = 11\sqrt{3} + 5\sqrt{3} + 8\sqrt{3}$

$\quad = 24\sqrt{3}$

Resultant force $= \sqrt{6^2 + (24\sqrt{3})^2}$

$\qquad\qquad\qquad = 42$ N

Direction $= \tan^{-1}\left(\dfrac{24\sqrt{3}}{6}\right)$

$\qquad\qquad = 81.8°$ to the direction of the original 3 N force.

③

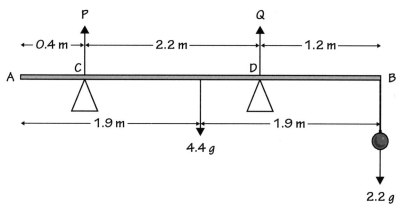

Redraw the diagram adding the forces and distances.

Taking moments about C, we obtain:

$4.4g \times 1.5 + 2.2g \times 3.4 = Q \times 2.2$

$\qquad\qquad 14.08 \times 9.8 = 2.2Q$

$\qquad\qquad\qquad\qquad Q = 62.72$ N

Taking moments about D, we obtain:

$P \times 2.2 + 2.2g \times 1.2 = 4.4g \times 0.7$

$\qquad\qquad\qquad\qquad P = 1.96$ N

④ (a) Tension in OW = 200 N (i.e. equal to the weight W).

(b)

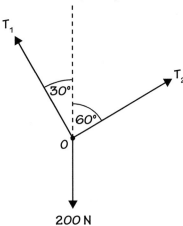

200 N

(c) Let tension in string OA = T_1 and tension in string OB = T_2

Resolving vertically, we obtain:

$$T_1 \cos 30° + T_2 \cos 60° = 200$$

$$\frac{\sqrt{3}}{2} T_1 + \frac{1}{2} T_2 = 200$$

$$\sqrt{3} T_1 + T_2 = 400 \qquad (1)$$

Resolving horizontally, we obtain:

$$T_1 \sin 30° = T_2 \sin 60°$$

$$\frac{1}{2} T_1 = \frac{\sqrt{3}}{2} T_2$$

$$T_1 = \sqrt{3} T_2 \qquad (2)$$

Substituting $T_1 = \sqrt{3} T_2$ into equation (1), we obtain:

$$\sqrt{3} \times \sqrt{3} T_2 + T_2 = 400$$

$$3T_2 + T_2 = 400$$

$$4T_2 = 400$$

$$T_2 = 100 \text{ N}$$

Substituting this value of T_2 into equation (2), we obtain:

$$T_1 = \sqrt{3} \times 100$$

$$= 173.21 \text{ N}$$

5 (a)

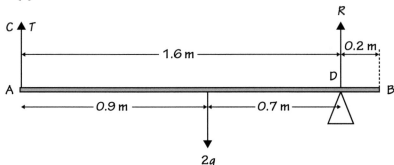

Taking moments about point D, we obtain:

$T \times 1.6 = 2g \times 0.7$

$T = \dfrac{2g \times 0.7}{1.6}$

$= 8.575$

$= 8.6\text{ N}$

(b) Resolving vertically, we obtain:

$T + R = 2g$

$8.6 + R = 2 \times 9.8$

$R = 11\text{ N}$

6 Resultant force in the direction of 18 N force:

$= 18 - 8\sqrt{3}\sin 30° - 9\sqrt{2}\sin 45°$

$= 2.0718\text{ N}$

Here we will take to upwards as the positive direction.

Resultant force in the direction at right-angles to the 18 N force:

$= 9\sqrt{2}\cos 45° - 8\sqrt{3}\cos 30°$

Here we will take to the right as the positive direction.

$= 3\text{ N}$

$R = \sqrt{2.0718^2 + (-3)^2}$

$= 3.646\text{ N}$

As this is negative the direction is opposite to the direction we have taken as positive. So this resultant acts to the left.

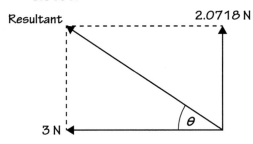

$\theta = \tan^{-1}\left(\dfrac{2.0718}{3}\right) = 34.6°$ (in the direction shown)

Or $90 - 34.6 = 55.4°$ to the left of the direction of the 18 N force.

Exam practice answers

Topic 15

1 (a)

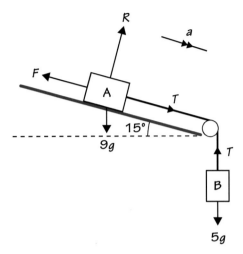

$F = \mu R$

  $= 0$ as $\mu = 0$.

Applying Newton's 2nd law parallel to the slope, we obtain:

$ma = T + mg \sin 15°$

  $9a = T + 9g \sin 15°$ (1)

Applying Newton's 2nd law to object B, we obtain:

$ma = 5g - T$

  $5a = 5g - T$ (2)

Adding equations (1) and (2) we obtain:

$14a = 5g + 9g \sin 15°$

  $a = 5.13 \text{ ms}^{-2}$

Substituting for a into equation (2) we obtain:

$5 \times 5.13 = 5g - T$

  $T = 23.35 \text{ N}$

(b) Forces acting down the slope $= 9g \sin 15° + 5g$

  $= 71.83 \text{ N}$

As the arrangement remains at rest, the minimum frictional force needed $= 71.83 \text{ N}$

Resolving perpendicular to the slope, we obtain:

$R = 9g \cos 15°$

  $= 85.19 \text{ N}$

$\mu = \dfrac{F}{R}$

  $= \dfrac{71.83}{85.19}$

  $= 0.84 \text{ (2 d.p.)}$

2 (a) $v = u + at$

    $15 = 0 + 50a$

      $a = 0.3 \text{ ms}^{-2}$

(b) Applying Newton's 2nd law of motion to the car, we obtain:
$$ma = 300 - F$$
$$800 \times 0.3 = 300 - F$$
$$F = 60 \text{ N}$$

(c) $v^2 = u^2 + 2as$
$$0 = 15^2 + 2a \times 500$$
$$a = -0.225 \text{ ms}^{-2}$$
$$F = ma$$
$$= 800 \times 0.225$$
$$= 180 \text{ N}$$

The minus sign tells us that it is a deceleration. As only the magnitude of the force is required we put the acceleration into the force equation as a positive value.

3 (a)

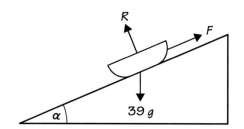

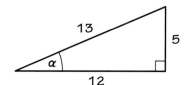

$$\tan \alpha = \frac{5}{12}$$

$$\sin \alpha = \frac{5}{13}$$

$$\cos \alpha = \frac{12}{13}$$

Resolving forces at right-angles to the slope, we obtain:
$$R = 39g \cos \alpha$$
$$= 39 \times 9.8 \times \frac{12}{13}$$
$$= 352.8 \text{ N}$$
$$F = \mu R$$
$$= 0.3 \times 352.8$$
$$= 105.84 \text{ N}$$

Applying Newton's 2nd law to the motion down the slope, we obtain:
$$ma = 39g \sin \alpha - F$$
$$39a = 39 \times 9.8 \times \frac{5}{13} - 105.84$$
$$a = 1.06 \text{ ms}^{-2}$$

(b)

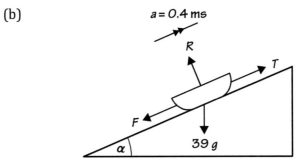

Applying Newton's 2nd law to the motion up the slope, we obtain:

$$ma = T - F - 39g \sin \alpha$$

$$39 \times 0.4 = T - 105.84 - 39 \times 9.8 \times \frac{5}{13}$$

$$T = 268.44 \text{ N}$$

4

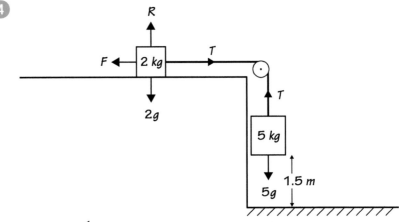

$$s = ut + \frac{1}{2}at^2$$

$$1.5 = 0 + \frac{1}{2} \times a \times 0.7^2$$

$$a = 6.122 \text{ ms}^{-2}$$

Applying Newton's 2nd law to the 5 kg mass, we obtain:

$$ma = 5g - T$$

$$T = 5g - 5a$$
$$= 5 \times 9.8 - 5 \times 6.122$$
$$= 18.39 \text{ N}$$

Applying Newton's 2nd law to the 2 kg mass, we obtain:

$$ma = T - F$$

$$2 \times 6.122 = 18.39 - F$$

$$F = 6.146 \text{ N}$$
$$R = mg = 2 \times 9.8 = 19.6 \text{ N}$$
$$\mu = \frac{F}{R} = \frac{6.146}{19.6} = 0.3 \text{ (1 d.p.)}$$

5 (a)

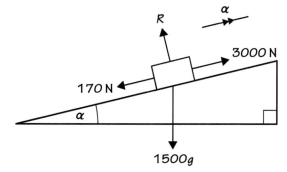

Applying Newton's 2nd law to the motion parallel to the slope, we obtain:

$$ma = 3000 - 1500g \sin \alpha - 170$$

$$1500a = 3000 - 1500 \times 9.8 \times \frac{1}{14} - 170$$

$$a = 1.9 \text{ ms}^{-2} \text{ (2 s.f.)}$$

(b) Resolving forces perpendicular to the slope, we obtain:

$$R = 1500g \cos \alpha$$
$$= 1500 \times 9.8 \times 0.9974$$
$$= 14662$$
$$= 15\,000 \text{ N (2 s.f.)}$$

$$\mu = \frac{F}{R} = \frac{170}{14662} = 0.01159 = 0.012 \text{ (2 s.f.)}$$

If $\sin \alpha = \frac{1}{14}$ then $\cos \alpha$
$$= \frac{1}{\sqrt{14^2 - 1^2}} = 0.9974$$

Topic 16

1 (a)

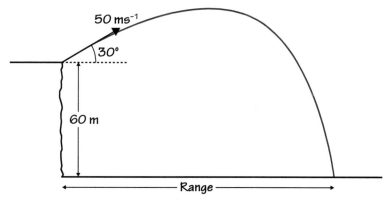

Considering the vertical motion of the ball and taking the upward direction as positive:

$$s = ut + \frac{1}{2}at^2$$

$$-60 = 50 \sin 30°t - 4.9t^2$$

$$4.9t^2 - 25t - 60 = 0$$

$$t = \frac{-b \pm \sqrt{b^2 - 4ac}}{2a}$$

$$= \frac{25 \pm \sqrt{(-25)^2 - 4(4.9)(-60)}}{2(4.9)}$$

$$t = 6.88 \text{ s}$$

Note the displacement is −60 as it is 60 m from the point of projection in the opposite direction.

Note there is also a negative solution which is ignored.

(b) Range = horizontal component of velocity × time

= 50 cos 30° × 6.88

= 298 m

2 (a) First we draw a right-angled triangle showing the angle θ and the known sides.

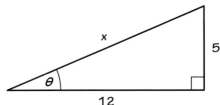

The length of the hypotenuse, x, can be found using Pythagoras' theorem.

$x = 13$

We now know $\cos \theta = \dfrac{12}{13}$ and $\sin \theta = \dfrac{5}{13}$

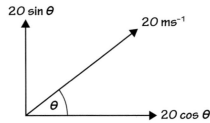

Initial horizontal component of velocity = 20 cos θ

$$= 20 \times \frac{12}{13} = 18.46 \text{ ms}^{-1}$$

Initial vertical component of velocity = 20 sin θ

$$= 20 \times \frac{5}{13} = 7.69 \text{ ms}^{-1}$$

(b) Initial velocity = 18.46**i** + 7.69**j**

3 (a) (i) $\sin \alpha = 0.8 = \dfrac{4}{5}$ so $\cos \alpha = \dfrac{3}{5}$

Initial vertical velocity, $u = 24.5 \sin \alpha = 24.5 \times 0.8$

$= 19.6 \text{ ms}^{-1}$

Considering the vertical motion taking the upward direction as positive, we have:

$u = 19.6$, $a = -9.8$, $s = 14.7$

Using $s = ut + \dfrac{1}{2} at^2$

$$14.7 = 19.6t - 4.9t^2$$
$$4.9t^2 - 19.6t + 14.7 = 0$$
$$t^2 - 4t + 3 = 0$$
$$(t - 1)(t - 3) = 0$$
$$t = 1 \text{ s or } 3 \text{ s}$$

Time it reaches the top of the first tree = 1 s

(ii) Horizontal component of the velocity = 24.5 cos ∝

$$= 24.5 \times \frac{3}{5}$$

$$= 14.7 \text{ ms}^{-1}$$

Time for the ball to travel between the two trees =

3 – 1 = 2 s

Distance = speed × time = 14.7 × 2 = 29.4 m

(b) Horizontal component of velocity is constant at 14.7 ms⁻¹

For vertical velocity after 0.75 s and taking the upward direction as positive, we have:

$u = 19.6$, $a = -9.8$, $t = 0.75$

$v = u + at$

$= 19.6 - 9.8 \times 0.75$

$= 12.25 \text{ ms}^{-1}$

Magnitude of velocity = $\sqrt{14.7^2 + 12.25^2}$

$$= 19.14 \text{ ms}^{-1}$$

$$\theta = \tan^{-1}\left(\frac{12.25}{14.7}\right)$$

$$= 39.8° \text{ (to the horizontal)}$$

  (a)

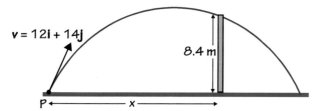

$v = 12i + 14j$

8.4 m

P ← x →

Let x = distance from P to the wall.

Using the equation of motion, $\mathbf{s} = \mathbf{u}t + \frac{1}{2}\mathbf{a}t^2$ we can write the following equation in terms of the vectors:

$\mathbf{s} = \mathbf{u}t + \frac{1}{2}\mathbf{a}t^2$

$$x\mathbf{i} + 8.4\mathbf{j} = (12\mathbf{i} + 14\mathbf{j})t - \frac{1}{2} \times 9.8\mathbf{j}t^2$$

Now looking at the motion in the vertical direction we have

$$8.4\mathbf{j} = 14\mathbf{j}t - 4.9t^2\mathbf{j}$$

Hence $8.4 = 14t - 4.9t^2$

$4.9t^2 - 14t + 8.4 = 0$

Notice that this quadratic can be divided by 0.7 so all the numbers in the equation become whole numbers.

$7t^2 - 20t + 12 = 0$

$(7t - 6)(t - 2) = 0$

$$t = \frac{6}{7}\text{s or 2 s}$$

$$t = 2 \text{ s}$$

The horizontal velocity is a constant 12 ms⁻¹.

Distance x = speed × time = 12 × 2 = 24 m

Horizontal distance of the wall = 24 m

Note that the $\frac{6}{7}$s is the time when the particle is at a height of 8.4 m on the way up. We use $t = 2$ s as this is when the particle is at a height of 8.4m on the way down.

(b) Horizontal velocity stays constant at 12 ms^{-1}.

$\mathbf{v} = \mathbf{u} + \mathbf{a}t$

$\qquad = (12\mathbf{i} + 14\mathbf{j}) - 9.8\mathbf{j}t$

When above the wall, $t = 2$ s.

Hence, $\mathbf{v} = 12\mathbf{i} + 14\mathbf{j} - 19.6\mathbf{j}$

$\qquad = 12\mathbf{i} - 5.6\mathbf{j}$

Speed $= |\mathbf{v}| = \sqrt{12^2 + (-5.6)^2}$

$\qquad\qquad = 13.24$ ms^{-1}

$$\theta = \tan^{-1}\left(\frac{5.6}{12}\right)$$

$\qquad = 25°$ (measured downwards from the horizontal)

5 (a) Using $\mathbf{s} = \mathbf{u}t + \frac{1}{2}\mathbf{a}t^2$

$(x\mathbf{i} + 0\mathbf{j}) = (21\mathbf{i} + 12\mathbf{j})t + \frac{1}{2}(-9.8\mathbf{j})t^2$

Considering just the vertical motion, the vertical displacement is 0.

$0\mathbf{j} = 12\mathbf{j}t - 4.9\mathbf{j}t^2$

$0 = 12t - 4.9t^2$

$0 = t(12 - 4.9t)$

Either $t = 0$ or $t = \dfrac{12}{4.9}$ s

Time of flight = 2.45 s

Considering just the horizontal motion, the horizontal displacement (i.e. the range) is x m.

$x\mathbf{i} = 21\mathbf{i}t$

Now, $t = 2.45$ s

$\qquad x\mathbf{i} = 21 \times 2.45\mathbf{i}$

$\qquad x = 51.45$ m

Horizontal range = 51.45 m

(b) At the maximum height, the vertical component of the velocity is 0.

Using $v^2 = u^2 + 2as$

$\qquad 0^2\mathbf{j} = 12^2\mathbf{j} - 2(9.8)s\mathbf{j}$

$\qquad s = 7.35$ m

Maximum height reached = 7.35 m

(c) When the particle hits the ground $t = 2.45$s

Using $v = u + at$

$\mathbf{v} = 21\mathbf{i} + 12\mathbf{j} + (-9.8\mathbf{j})(2.45)$

$\qquad = 21\mathbf{i} + 12\mathbf{j} - 24.01\mathbf{j}$

$\qquad = 21\mathbf{i} - 12.01\mathbf{j}$

Speed $= |\mathbf{v}| = \sqrt{21^2 + (-12.01)^2}$

$\qquad\qquad = 24.19$ ms^{-1}

$$\theta = \tan^{-1}\left(\frac{12.01}{21}\right)$$

$\qquad = 29.77°$

The particle is travelling at an angle of 29.77° down from the horizontal.

6 (a)

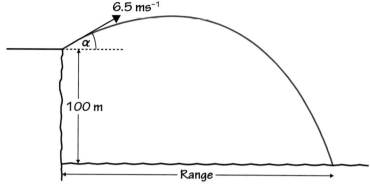

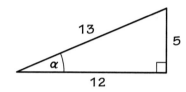

$$\tan \alpha = \frac{5}{12}$$
$$\sin \alpha = \frac{5}{13}$$
$$\cos \alpha = \frac{12}{13}$$

When the stone enters the sea, the vertical displacement is −100.

Considering the vertical motion and using $s = ut + \frac{1}{2}at^2$, we have:

$$-100 = 6.5 \sin \alpha \, t + \frac{1}{2}(-9.8)t^2$$

Now $\sin \alpha = \frac{5}{13}$ so

$$-100 = 6.5 \times \frac{5}{13}t - 4.9t^2$$

$$4.9t^2 - 2.5t - 100 = 0$$

$$t = \frac{-b \pm \sqrt{b^2 - 4ac}}{2a} = \frac{2.5 \pm \sqrt{(-2.5)^2 - 4(4.9)(-100)}}{2(4.9)} = 4.78 \text{ s}$$

(b) Range = horizontal component of velocity × time
    $$= 6.5 \cos \alpha \times 4.78$$
    $$= 28.68 \text{ m}$$

> The negative time is ignored.

(c) Considering the vertical motion:
    $$v^2 = u^2 + 2as$$
    $$= 2.5^2 + 2(-9.8)(-100)$$
    $$= \pm 44.34$$

Horizontal component of velocity $= 6.5 \cos \alpha = 6.5 \times \frac{12}{13} = 6$

Speed $= \sqrt{6^2 + (-44.34)^2}$
    $$= 44.74 \text{ ms}^{-1}$$

Angle to the horizontal $= \tan^{-1}\left(\frac{44.34}{6}\right)$
    $$= 82.29° \text{ (below the horizontal)}$$

Exam practice answers

Remember to include the negative sign as the mass m is decreasing with time. Always ask yourself if the quantity in the question is increasing or decreasing with time.

You often have to use the rules of logarithms in differential equation questions.

1 (a) $\dfrac{dm}{dt} \propto -m$

$\dfrac{dm}{dt} = -km$

Separating variables and integrating, we obtain:

$\int \dfrac{dm}{m} = -k \int dt$

$\ln m = -kt + c$

When $t = 0$, $m = 4$ so $\ln 4 = 0 + c$

Hence, $\ln m = -kt + \ln 4$

$\ln m - \ln 4 = -kt$

$\ln \left(\dfrac{m}{4}\right) = -kt$

Taking exponentials of both sides, we obtain:

$\dfrac{m}{4} = e^{-kt}$

$m = 4e^{-kt}$

When $t = 10$, $m = 2$ so $2 = 4e^{-10k}$

$0.5 = e^{-10k}$

Taking ln of both sides:

$\ln 0.5 = -10k$

$k = 0.0693$

Hence $m = 4e^{-0.0693t}$

When $t = 30$, $m = 4e^{-0.0693 \times 30} = 0.5$ mg

(b) $m = 4e^{-0.0693t}$

$0.3 = 4e^{-0.0693t}$

$0.075 = e^{-0.0693t}$

$\ln 0.075 = -0.0693t$

$t = 37$ days (nearest day)

Watch out

This an indefinite integral, so a constant of integration c must be included.

2 (a) $\dfrac{dm}{dt} \propto -m$

$\dfrac{dm}{dt} = -km$

(b) Separating variables and integrating, we obtain

$\int \dfrac{dm}{m} = -k \int dt$

$\ln m = -kt + c$

When $t = 0$, $m = m_0$, so $\ln m_0 = 0 + c$ giving $c = \ln m_0$

Hence $\ln m = -kt + \ln m_0$

$\ln m - \ln m_0 = -kt$

$\ln \left(\dfrac{m}{m_0}\right) = -kt$

Taking exponentials of both sides, we obtain:

$$\frac{m}{m_0} = e^{-kt}$$

$$m = m_0\, e^{-kt}$$

(c) (i)

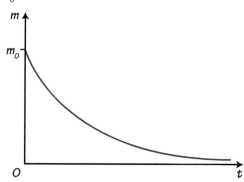

(ii) When t becomes large, m becomes small. Since $m + M = A$, as m decreases M increases so as to keep A constant. M will approach (but never reach) the value of A.

(iii) Rate of decrease of m = rate of increase of M

Hence, $-\dfrac{dm}{dt} = \dfrac{dM}{dt}$

Now $\dfrac{dm}{dt} = -km$ so $-\dfrac{dm}{dt} = km$

So $\dfrac{dM}{dt} = km$ but as $m = A - M$

$$\frac{dM}{dt} = k(A - M)$$

3 (a) $\dfrac{d\theta}{dt} \propto -(\theta - 15)$

$$\frac{d\theta}{dt} = -k(\theta - 15)$$

Separating variables and integrating, we obtain:

$$\int \frac{d\theta}{\theta - 15} = -k\!\int dt$$

$\ln(\theta - 15) = -kt + c$

(b) (i) When $t = 0$ min, $\theta = 100\,^{\circ}\text{C}$

Hence, $\ln(100 - 15) = 0 + c$

$c = \ln 85$

(ii) $\qquad \ln(\theta - 15) = -kt + \ln 85$

$\ln(\theta - 15) - \ln 85 = -kt$

$\ln\left(\dfrac{\theta - 15}{85}\right) = -kt$

Taking exponentials of both sides:

$$\frac{\theta - 15}{85} = e^{-kt}$$

$\theta - 15 = 85e^{-kt}$

$\theta = 85e^{-kt} + 15$

Remember that decreasing rates always have a negative sign.

Remember that the temperature decreases with time, so a negative sign is needed.

(iii) When $t = 10$, $\theta = 50\,^\circ C$.

$$50 = 85e^{-10k} + 15$$
$$35 = 85e^{-10k}$$
$$\frac{35}{85} = e^{-10k}$$

Taking ln of both sides, we obtain:

$$\ln\left(\frac{35}{85}\right) = -10k$$
$$k = 0.089$$

(c) $\theta = 85e^{-0.089t} + 15$

When $\theta = 70\,^\circ C$

$$70 = 85e^{-0.089t} + 15$$
$$55 = 85e^{-0.089t}$$
$$\frac{55}{85} = e^{-0.089t}$$

Taking ln of both sides

$$\ln\left(\frac{55}{85}\right) = -0.089t$$
$$t = 4.89 \text{ min}$$

4 $\quad \dfrac{dx}{dt} = \dfrac{kx(n - x)}{n}$

Separating variables and integrating, we obtain:

$$\int \frac{1\,dx}{x(n - x)} = \frac{k}{n}\int dt$$

Changing $\dfrac{1}{x(n - x)}$ to partial fractions, we have

$$\frac{1}{x(n - x)} = \frac{A}{x} + \frac{B}{n - x}$$
$$1 = A(n - x) + Bx$$

Let $x = n$, so $1 = A(n - n) + Bn$, giving $B = \dfrac{1}{n}$

Let $x = 0$, so $1 = An + 0$, giving $A = \dfrac{1}{n}$

Hence $\dfrac{1}{x(n - x)} = \dfrac{1}{nx} + \dfrac{1}{n(n - x)}$

Now $\displaystyle\int\left(\frac{1}{nx} + \frac{1}{n(n - x)}\right) dx = \frac{k}{n}\int dt$

$$\frac{1}{n}\int\left(\frac{1}{x} + \frac{1}{(n - x)}\right) dx = \frac{k}{n}\int dt$$

$$\int\left(-\frac{1}{x} - \frac{1}{(n - x)}\right) dx = -k\int dt$$

$$-\ln x + \ln(n - x) = -kt + \ln A$$

$$\ln\left(\frac{n - x}{x}\right) = -kt + \ln A$$

$$\ln\left(\frac{n - x}{x}\right) - \ln A = -kt$$

$$\ln\left(\frac{n - x}{Ax}\right) = -kt$$

You need change $\dfrac{1}{x(n - x)}$ to partial fractions before integrating it. Remember you may have to use techniques learned in Pure maths to integrate expressions.

Multiplying both sides by $-n$.

Notice the $-kt$ in the answer so we multiply both sides by -1.

Rather than use the constant of integration as c we can use ln A so that we can combine the ln terms.

Taking exponentials of both sides, we obtain:

$$\frac{n - x}{Ax} = e^{-kt}$$

$$\frac{n - x}{x} = Ae^{-kt}$$

$$n - x = xAe^{-kt}$$

$$n = x + xAe^{-kt}$$

$$n = x(1 + Ae^{-kt})$$

$$x = \frac{n}{(1 + Ae^{-kt})}$$

5 (a) $\dfrac{dx}{dt} + x = 2$

$$\frac{dx}{dt} = 2 - x$$

Separating variables and integrating, we obtain:

$$\int \frac{1}{2 - x}\, dx = \int dt$$

$$-\int \frac{-1}{2 - x}\, dx = t + c$$

$$-\ln|2 - x| = t + c$$

When $t = 0, x = 0$

$$c = -\ln 2$$

Hence, $-\ln|2 - x| = t - \ln 2$

$$\ln 2 - \ln|2 - x| = t$$

$$t = \ln\left(\frac{2}{|2 - x|}\right)$$

When $x = 1$, $t = \ln\left(\dfrac{2}{|2 - x|}\right) = \ln 2 = 0.693$ s

(b) $t = \ln\left(\dfrac{2}{|2 - x|}\right)$

Taking exponentials of both sides, we obtain:

$$\frac{2}{2 - x} = e^t$$

$$2 = (2 - x)e^t$$

$$\frac{2}{e^t} = 2 - x$$

$$x = 2 - \frac{2}{e^t}$$

$$= 2 - 2e^{-t}$$

$$= 2(1 - e^{-t})$$

$$a = \frac{d^2x}{dt^2}$$

$$\frac{dx}{dt} = 2e^{-t}$$

$$\frac{d^2x}{dt^2} = -2e^{-t}$$

So $a = -2e^{-t}$

6 (a) The only force acting on the box in the horizontal direction, is the resistive force which acts in the opposite direction to the velocity of the box.
Hence, we can apply Newton's 2nd law to give:

$$m\frac{dv}{dt} = -0.4v^2$$

$$2\frac{dv}{dt} = -0.4v^2$$

$$2\frac{dv}{dt} + 0.4v^2 = 0$$

Dividing both sides of the equation by 0.4, we obtain:

$$5\frac{dv}{dt} + v^2 = 0$$

(b) $$5\frac{dv}{dt} = -v^2$$

Separating variables and integrating, we obtain:

$$5\int\frac{dv}{v^2} = -\int dt$$

$$5\int v^{-2}\, dv = -\int dt$$

$$\frac{5v^{-1}}{-1} = -t + c$$

$$-\frac{5}{v} = -t + c$$

When $t = 0$, $v = 5$, so:

$$-\frac{5}{5} = 0 + c, \text{ giving } c = -1$$

$$-\frac{5}{v} = -t - 1$$

Multiplying through by −1 we obtain:

$$\frac{5}{v} = t + 1$$

$$v = \frac{5}{t + 1}$$

(c) Using the above equation, if t becomes very large, the fraction will approach but never reach zero. This means the box will never stop which is impossible when friction opposes the motion.